TI-83 Plus, TI-84 Plus, and TI-89 Manual

KATHLEEN MCLAUGHLIN
University of Connecticut

DOROTHY WAKEFIELD
University of Connecticut Health Center

to accompany

THE SULLIVAN STATISTICS SERIES:

Statistics: Informed Decisions Using Data, Third Edition

Fundamentals of Statistics, Third Edition

Michael Sullivan, III
Joliet Junior College

Prentice Hall
is an imprint of

Reproduced by Pearson Prentice Hall from electronic files supplied by the author.

Publishing as Pearson Prentice Hall, Upper Saddle River, NJ 07458.

Printed in the United States of America.

ISBN-13: 978-0-321-57778-8
ISBN-10: 0-321-57778-7

1 2 3 4 5 6 OPM 10 09 08

Prentice Hall
is an imprint of

www.pearsonhighered.com

Introduction

A *Graphing Calculator Manual for the TI-83Plus, TI-84 Plus and TI-89* is one of a series of companion technology manuals that provide hands-on technology assistance to users of Sullivan *Statistics: Informed Decisions Using Data,* Third edition.

Detailed instructions for working selected examples and problems from *Statistics: Informed Decisions Using Data* are provided in this manual. To make the correlation with the text as seamless as possible, the table of contents includes page references for both the Sullivan text and this manual.

Contents:

Getting Started with the TI-84 and TI-83 Graphing Calculators

▸ Overview

This manual is designed to be used with the TI-84 and TI-83 families of Graphing Calculators. These calculators have a variety of useful functions for doing statistical calculations and for creating statistical plots. The commands for using the statistical functions are basically the same for the TI-84's and TI-83's. All TI-84 calculators, the TI-83 Plus Calculator and the TI-83 Silver Edition can receive a variety of software applications that are available through the TI website (www.education.ti.com). TI also will provide downloadable updates to the operating systems of these calculators. These features are not available on the TI-83.

All of the data sets referenced in this manual may be found on the data disk packaged in the back of every new copy of Sullivan's *Statistics: Informed Decisions Using Data*, Third Edition. If needed, the data sets may also be downloaded from the Pearson Statistics Data Sets Website at www.pearsonhighered.com/datasets. The requirements for the transfer of data from a computer differ from one calculator to another. Some TI calculators are sold with the necessary connector. For those calculators that do not come with a connector, you can purchase the connector at the TI website: www.education.ti.com. (Note: In order to do examples in this manual, you can simply enter the data values for each example directly into your calculator. It is not necessary to use the graph link to download the data into your calculator. The download procedure using the computer link is an optional way of entering data.)

Throughout this manual all instructions and screen shots use the TI-84. These instructions and screen shots are also compatible with the TI-83 calculators.

Before you begin using the TI-84 or TI-83 calculator, spend a few minutes becoming familiar with its basic operations. First, notice the different colored keys on the calculator. On the TI-84's, the white keys are the number keys; the light gray keys on the right are the basic mathematical functions; the dark gray keys on the left are additional mathematical functions; the remaining dark gray keys are the advanced functions; the light gray keys just below the viewing screen are used to set up and display graphs, and the light gray arrow keys are used for moving the cursor around the viewing screen. On the TI-83's, the white keys are the number keys; the blue keys on the right are the basic mathematical functions; the dark gray keys on the left are additional mathematical functions; the remaining dark gray keys are the advanced functions; the blue keys just below the viewing screen are used to set up and display graphs, and the blue arrow keys are used for moving the cursor around the viewing screen.

The primary function of each key is printed in white on the key. For example, when you press **STAT**, the STAT MENU is displayed.

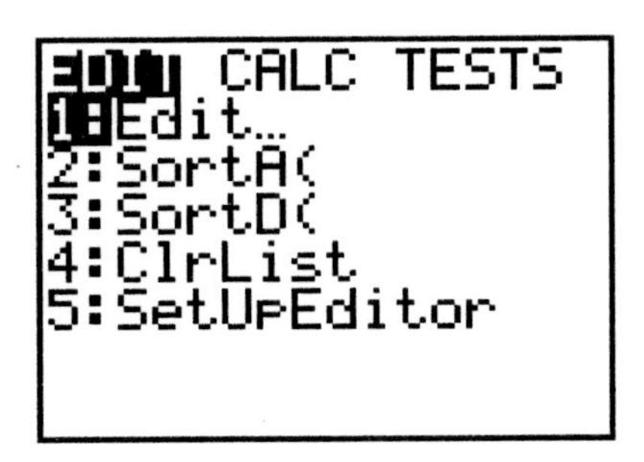

The secondary function of each key is printed in blue on the TI-84's (yellow on the TI-83's) above the key. When you press the **2nd** key (found in the upper left corner of the keys), the function printed above the key becomes active and the cursor changes from a solid rectangle to an ↑ (up-arrow). For example, when you press **2nd** and the x^2 key, the $\sqrt{\ }$ function is activated.
The notation used in this manual to indicate a secondary function is **'2nd'** followed by the name of the secondary function. For example, to use the LIST function, found above the **STAT** key, the notation used in this manual is **2nd [LIST]**. The LIST MENU will then be activated and displayed on the screen.

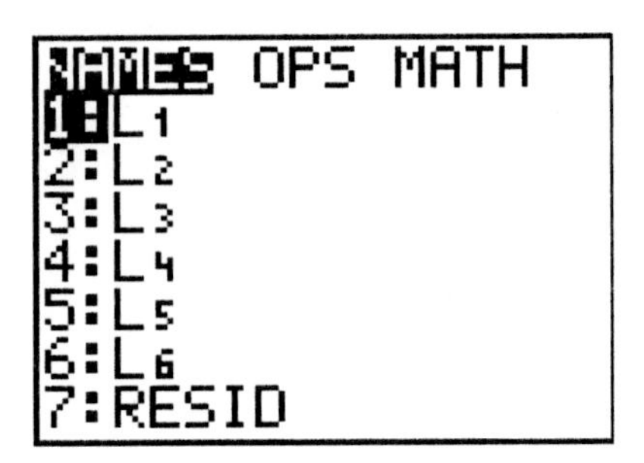

The alpha function of each key is printed in green above the key. When you press the green **ALPHA** key, the function printed in green above the key is activated and the cursor changes from a solid rectangle to **A**.

In this manual you will find detailed explanations of the different statistical functions that are programmed into the TI-84 and TI-83 graphing calculators. These explanations will accompany selected examples from your textbook. This will give you the opportunity to learn the various calculator functions as they apply to the specific statistical material in each chapter.

▸ Getting Started

To operate the calculator, press **ON** in the lower left corner of the calculator. Begin each example with a blank screen, with a rectangular cursor flashing in the upper left corner. If you turn on your calculator and you do not have a blank screen, press the **CLEAR** key. You may have to press **CLEAR** a second time in order to clear the screen. If using the **CLEAR** key does not clear the screen, you can push **2nd** **[QUIT]** (Note: **QUIT** is found above the **MODE** key.)

▸ Helpful Hints

To adjust the display contrast, push and release the 2nd key. Then push and hold the up arrow ▲ to darken or the down arrow ▼ to lighten.

The calculator has an automatic turn off that will turn the calculator off if it has been idle for several minutes. To restart, simply press the **ON** key.

There are several different graphing techniques available on the TI-84 and TI-83 calculators. If you inadvertently leave a graph on and attempt to use a different graphing function, your graph display may be cluttered with extraneous graphs, or you may get an ERROR message on the screen.

There are several items that you should check before graphing anything. First, press the **Y=** key, found in the upper left corner of the key pad, and clear all the Y-variables. The screen should look like the following display:

Plot1 Plot2 Plot3
\Y1=
\Y2=
\Y3=
\Y4=
\Y5=
\Y6=
\Y7=

If there are any functions stored in the Y-variables, simply move the cursor to the line that contains a function and press **CLEAR** **ENTER**.

Next, press 2nd **[STAT PLOT]** (found on the **Y=** key) and check to make sure that all the STAT PLOTS are turned **OFF**.

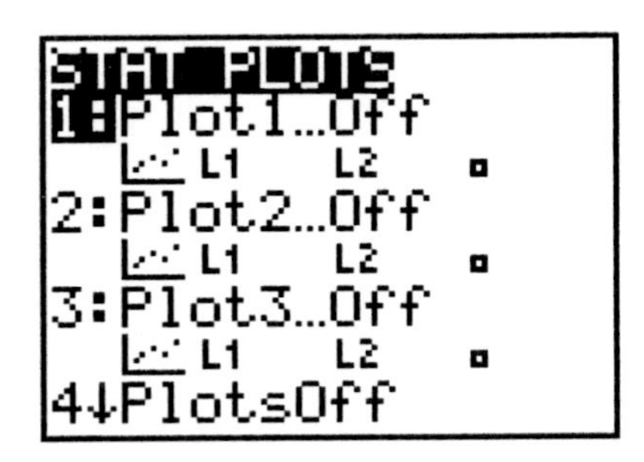

If you notice that a Plot is turned **ON**, select the Plot by using the down arrow key to highlight the number to the left of the Plot , press **ENTER** and move the cursor to **OFF** and press **ENTER**. Press **2nd [QUIT]** to return to the home screen.

Supplementary Information for the TI-89, the TI-Nspire and the TI-Nspire CAS

At the end of each chapter, there are instructions for the TI-89 Titanium Graphing Calculator and for the TI-Nspire Handhelds.

The TI-89 calculator is similar to the TI-84. The differences between the two calculators are mainly differences in how the various statistical and graphical menus are accessed. Once you have accessed a specific menu, the steps within the menu are similar. The TI-89 instructions at the end of each chapter explain how to access the menus that you need for the examples in that particular chapter and give an overview of the steps within the specific menu. If you need more detail on a specific example, refer to the TI-84 instructions for that example.

The TI-Nspire Handheld comes with two interchangeable keyboards: TI-Nspire keyboard and Ti-84 Plus keyboard. If you are using a TI-Nspire, you have two options. Option#1 is to use the TI-84 Plus keyboard and simply follow the TI-84 instructions in this manual. Option#2 is to use the TI-Nspire keyboard and follow the instructions at the end of each chapter.

The TI-Nspire CAS Handheld does not have the interchangeable TI-84 Plus keyboard. To use this handheld, follow the instructions for the TI-Nspire at the end of each chapter.

Data Collection

Section 1.2

▸ Example 3 (pg.26) **Generating a Simple Random Sample**

The first step is to set the *seed* by selecting any 'starting number' and storing this number in **rand**. Suppose, for this example, that we select the number '34' as the starting number. Type **34** into your calculator and press the **STO** key found in the lower left section of the calculator keys. Next press the **MATH** key found in the upper left section of the calculator keys. The Math Menu will appear.

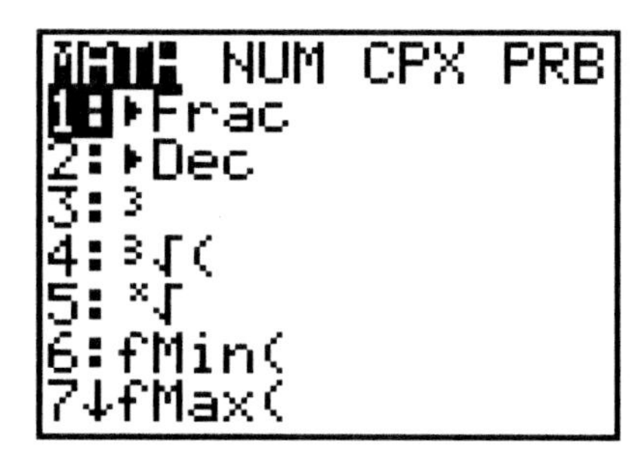

Use the right arrow key, ▶ found in the upper right section of the calculator keys, to move the cursor to highlight **PRB**. The Probability Menu will appear.

```
MATH NUM CPX PRB
1:rand
2:nPr
3:nCr
4:!
5:randInt(
6:randNorm(
7:randBin(
```

The first selection on the **PRB** menu is **rand,** which stands for 'random number'. Notice that this highlighted. Simply press **ENTER** twice and the starting value of '34' will be stored into **rand** and will be used as the *seed* for generating random numbers. (Note: This example uses '34' as the *seed,* but you can use any number as a seed for your random number generator.)

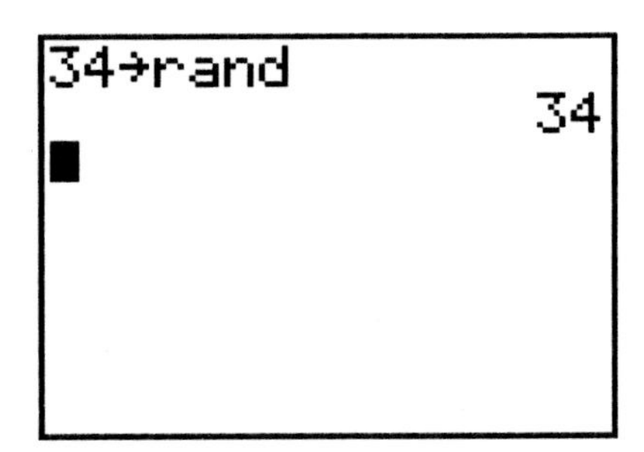

Now you are ready to generate a random integer. Press MATH again and the Math Menu will appear. Use the right arrow key, ▶, to move the cursor to highlight **PRB**. The Probability Menu will appear. Select **5:RandInt(** by using the down arrow key, ▼, to highlight it and pressing ENTER or by pressing the 5 key. **RandInt(** should appear on the screen. This function requires two values: the starting integer, followed by a comma (the comma is found on the black key above the 7 key), and the ending integer. To complete the **RandInt** command, close the parentheses and press ENTER. (Note: Closing the parenthesis at the end of the command is optional.) This command will generate one random number.

For this example, the starting integer is **1** and the ending integer is **30**. The random number generated in this example is **11**.

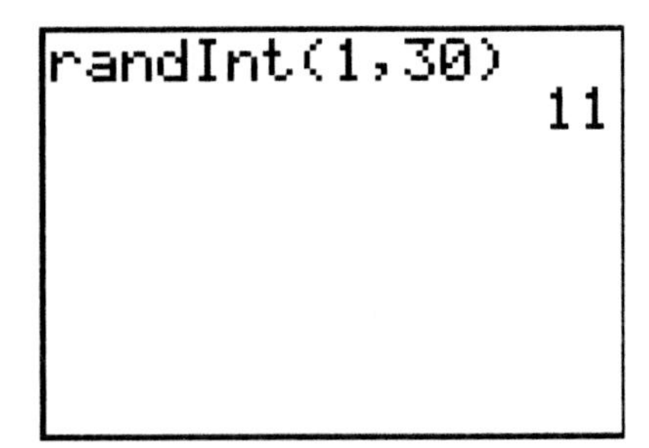

If you press ENTER again, a second random number between 1 and 30 will be generated. Continue pressing ENTER until you have generated 5 **distinct** random numbers. (Note: The TI-84 uses a method called "sampling with replacement" to generate random numbers. This means that it is possible to select the same integer twice.)

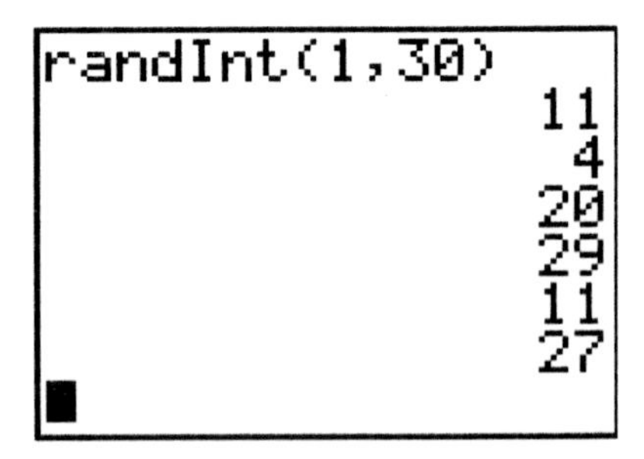

The five distinct random numbers generated are: 11, 4, 20, 29 and 27.

If you want to generate several random numbers with one command, you can change the **RandInt** command so that it contains three values: the starting value, the ending value and the number of values you want to generate. Since duplicates are possible, it is good practice to generate a few more numbers than are actually needed.

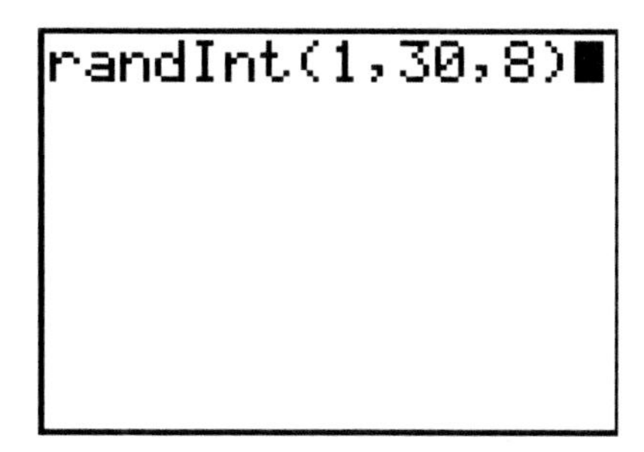

Press **ENTER** and a partial display of the 8 random integers should appear on your screen. (Note: your numbers will probably be different from the ones you see here. The numbers that are generated will depend on the *seed* that is initially selected.)

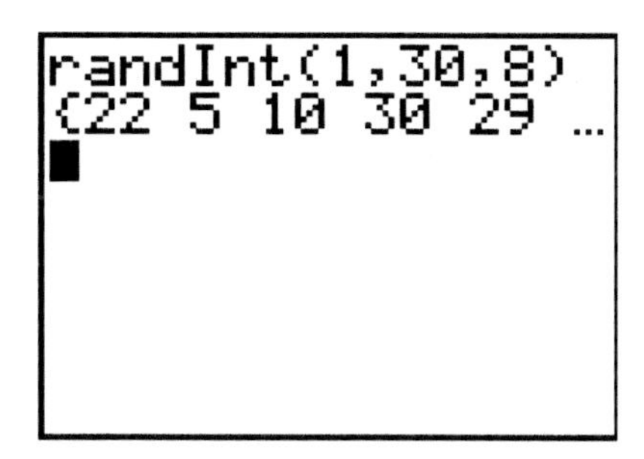

Use the right arrow to scroll through your 8 items and select the first five distinct random numbers.

▸ Problem 11 (pg.28) Generating a Simple Random Sample

a. To obtain a simple random sample of size 10, press MATH, use the right arrow to highlight **PRB** and select **5: randInt**. Enter the starting value of **1**, the ending value of **50** and a sample size of **15**. We are generating more numbers than we actually need because of the possibility of getting duplicates in the sample. To obtain your sample of 10, select the first 10 distinct numbers in the sample that you generated. (Note: In this example, we did not set a new seed. Setting a new seed every time that you generate a random sample is optional. It is not required.)

b. Repeat the steps in part a. and generate another sample of integers.

TI-89 Instructions

These instructions are designed to give you an overview of the method for generating random samples on the TI-89.

To set the *seed:*
Press the HOME key. Press 2nd and 5 for the **MATH** menu. Select **7:Probability** and select **6:RandSeed.** On the input line, next to **RandSeed**, type in any number and press ENTER.

Generating a random sample:
To generate a random integer between 1 and n (for example, a random integer between 1 and 30), from the HOME screen, press 2nd and 5 for the **MATH** menu. Select **7:Probability** and select **4:Rand(.** On the input line, next to **rand(**, type **30)** and press ENTER.

If you press ENTER again, a second random number between 1 and 30 will be generated. Continue pressing ENTER until you have generated as many distinct random numbers as you need. (Note: The TI-89 uses a method called "sampling with replacement" to generate random numbers. This means that it is possible to generate the same integer twice.)

TI-*n*spire Instructions:

These instructions are designed to give you an overview of the method for generating random samples on the **TI-*n*spire** handhelds.

Press [home] and select **6:New Document**. (Note: If you currently have a document open, the next screen will ask if you want to save the document. Press [tab] to select **No**. Press [enter].) Select **1:Add Calculator.**

To set the *seed:*
Press [menu], select **5:Probability,** select **4: Random,** and select **6:Seed.** Next to **RandSeed,** type in any number and press [enter].

Generating a random sample:
To generate a set of random integers between 1 and n (for example, a set of random integers between 1 and 30), press [menu], select **5:Probability,** select **4: Random,** and select **2:Integer.** On the input line next to **ranInt(** type in **1** followed by a comma, **30** followed by a comma and the number of integers you want to generate. Close the parenthesis and press [enter]. (Note: The **TI-*n*spire** uses a method called "sampling with replacement" to generate random numbers. This means that it is possible to generate the same integer twice.)

Organizing and Summarizing Data

Section 2.2

Example 2 (pg. 83) A Histogram for Discrete Data

To create a histogram, you have two choices: 1): enter all the individual data points from Table 8 on pg. 82 into one column or 2): enter the data values into one column and the frequencies into another column using the summarized data in Table 9 on pg. 83. For this example, we will use the summarized data.

To create this histogram, you must enter information into List1 (**L1**) and List 2 (**L2**) on your calculator. You will enter the 'number of customers' into **L1** and the frequencies into **L2**. Press **STAT** and the Statistics Menu will appear.

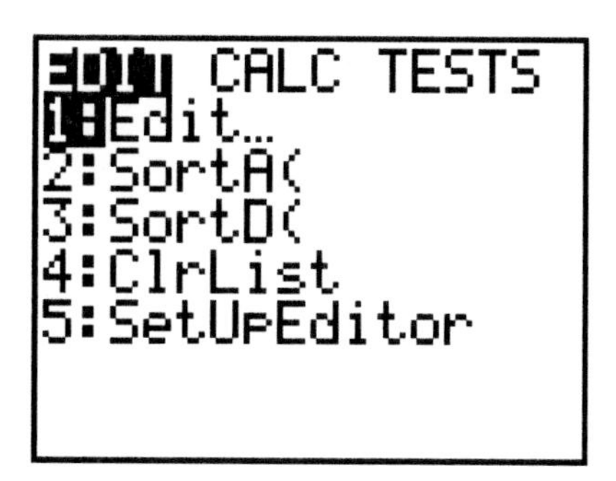

Press **ENTER** and lists **L1**, **L2** and **L3** will appear.

L1 L2 L3 2
------ ------ ------
L2(1)=

If the lists already contain data, you should clear them before beginning this example. Move your cursor so that the List name (**L1**, **L2**, or **L3**) of the list that contains data is highlighted.

L1	L2	L3 1
1	4	------
2	6	
3	8	
10		

L1 ={1,2,3,10}

Press **CLEAR** **ENTER**. Repeat this process until all three lists are empty.

L1	L2	L3 1
	4	------
	6	
	8	

L1(1)=

To enter the data values into **L1,** move your cursor so that it is positioned in the 1st position in **L1**. Type in the first value, **1,** and press **ENTER** or use the down arrow. Enter the next value, **2.** Continue this process until all 11 data values are entered into **L1**. Now use the up-arrow to scroll to the top of **L1**. As you scroll through the data, check it. If a data point is incorrect, simply move the cursor to highlight it and type in the correct value. When you have moved to the 1st value in **L1**, use the right arrow to move to the first position in **L2**. Enter the frequencies into **L2**.

L1	L2	L3 3
1	1	
2	6	
3	1	
4	4	
5	7	
6	11	
7	5	

L3(1)=

Before graphing the histogram, make sure that there are no functions in the Y-registers. To do this, press the **Y=** key. If there are any functions stored in any of the Y-values, simply move the cursor to the line that contains a function and press **CLEAR** . Now you are ready to graph the histogram. Press **2nd** **[STAT PLOT]** (located above the **Y=** key).

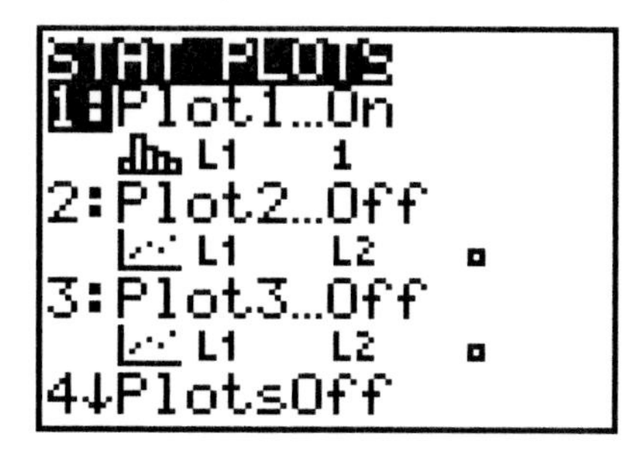

Select Plot1 by pressing **ENTER**.

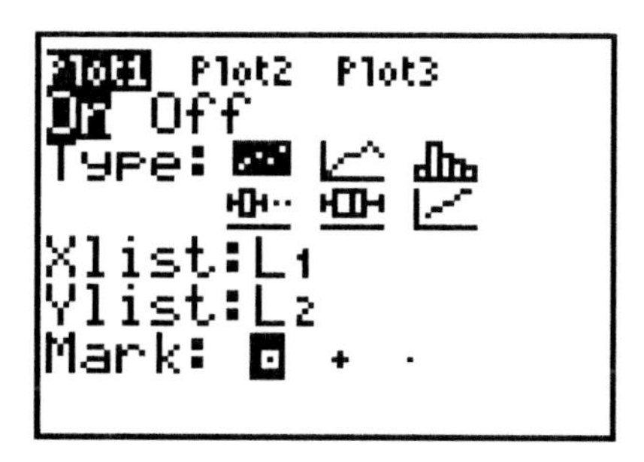

Notice that Plot1 is highlighted. On the next line, notice that the cursor is flashing on **ON** or **OFF**. Position the cursor to **ON** and press **ENTER** to select it. The next two lines on the screen show the different types of graphs. Move your cursor to the symbol for histogram (3rd item in the 1st line of **Type**) and press **ENTER**.

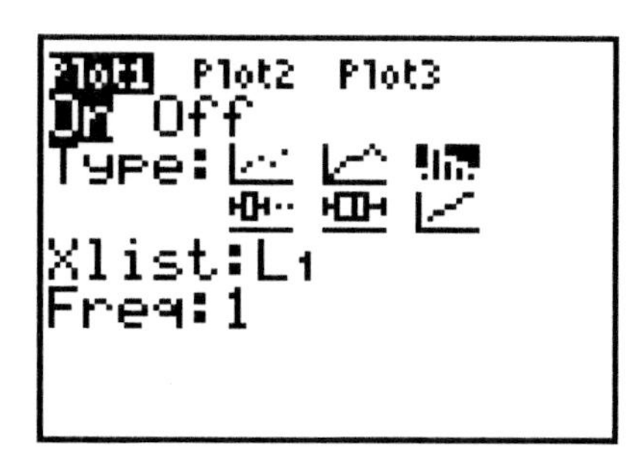

The next line is **Xlist**. Use the down arrow to move to this line. On this line, you indicate where the data values are stored. In most graphing situations, the data are entered into **L1,** so **L1** is the default option. Notice that the cursor is flashing on **L1**. Push **ENTER** to select **L1**. The last line is the frequency line. On this line '**1**' is the default. The cursor should be flashing on **1**. Change **1** to **L2** by pressing **2nd** **[L2]**. (Note: **L2** is found above the **2** key.)

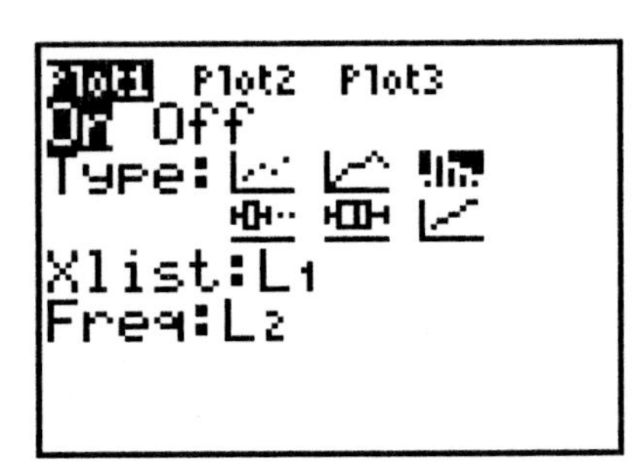

To view a histogram of the data, press ZOOM.

There are several options in the Zoom Menu. Using the down arrow, scroll down to option 9, **ZoomStat,** and press ENTER. A histogram should appear on the screen.

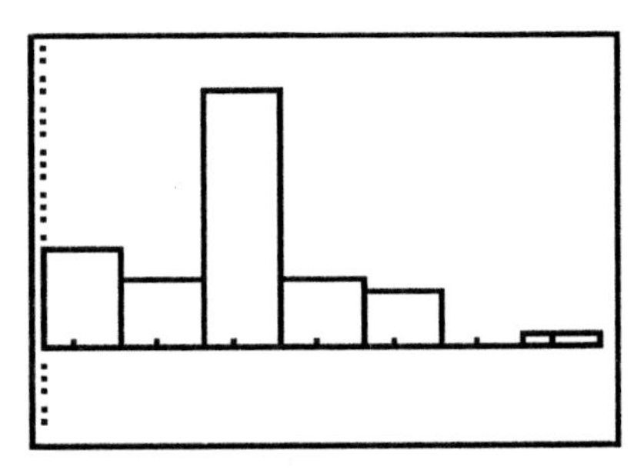

This histogram is not exactly the same as the frequency histogram in Fig. 7(a) on pg. 84 of your textbook. You can adjust the histogram so that it does look exactly like the one in your text. Press Window and set **Xmin** to 1, **Xmax** to 12 (this one extra data value is needed to complete the last bar of the histogram), and **Xscl** equal to 1, which is the difference between successive data values in the frequency distribution. Note: In many cases it is not necessary to change the values for **Ymin**, **Ymax** or **Yscl**. What you must do is to check these values and make sure that **Ymin** is a small negative value (a value between –6 and –1 would be good) and **Ymax** should be slightly larger than the largest frequency value in your dataset. A good value for Ymax for this example is 12. You never need to adjust **Yscl**.

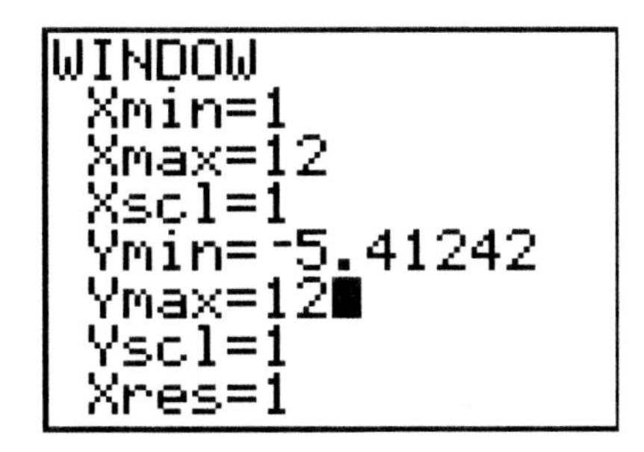

Press GRAPH.

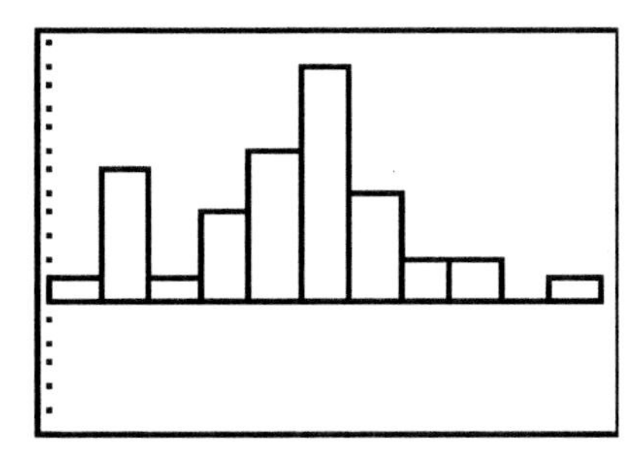

Notice the TRACE key. If you press it, a flashing cursor, *, will appear at the top of the 1st bar of the histogram.

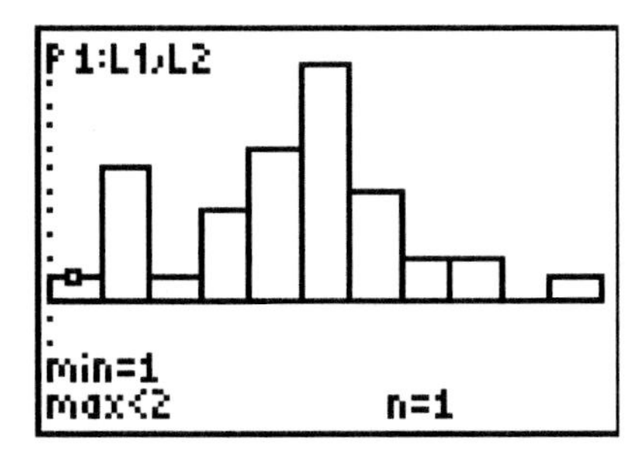

Notice the information at the bottom of the screen. **Min** is the actual data value for the first bar of the histogram. In this example, the first data value is **1**. We do not need to use the Max value in this example. "n=1" tells us that there is only one data point in the dataset that has a value of **1**. You can use the right arrow to move through each of the bars. For example, if you move to the 5th bar in the histogram, you will see that the data value for that bar is **5** and that there are 7 data points in the dataset that have a value of **5**.
Now that you have completed this example, turn Plot1 **OFF**. Using 2nd [STAT PLOT], select Plot1 by pressing ENTER and highlighting **OFF**. Press ENTER and 2nd [QUIT] (located above the [MODE] key.) (Note: Turning Plot1 **OFF** is optional. You can leave it ON but leaving it ON will effect other graphing operations of the calculator.)

To create the relative frequency histogram as shown in Fig. 7(b) on pg. 84, replace the frequencies in L2 with relative frequencies. One way to do this is to CLEAR L2 and enter the relative frequencies from Table 9 on pg. 83. Alternately, without clearing the frequencies, move the cursor so that it highlights the label, L2, at the top of the column and press ENTER. On the data entry line, type in 2nd L2/40. and press ENTER.

L1	L2	L3 2
1	1	------
2	6	
3	1	
4	4	
5	7	
6	11	
7	5	

L2 =L2/40

Press ENTER.

L1	L2	L3 2
1	.025	------
2	.15	
3	.025	
4	.1	
5	.175	
6	.275	
7	.125	

L2(1)=.025

Press Window to adjust the Graph Window. Set **Xmin** equal to 1 and **Xmax** equal to 12. Set **Xscl** equal to 1. Set **Ymin** to -.1 and **Ymax** to .28. Press GRAPH and the relative frequency histogram should appear.

You can press TRACE and scroll through the bars of the histogram. The minimum and maximum values of the class will appear. **n** is now the relative frequency of the class.

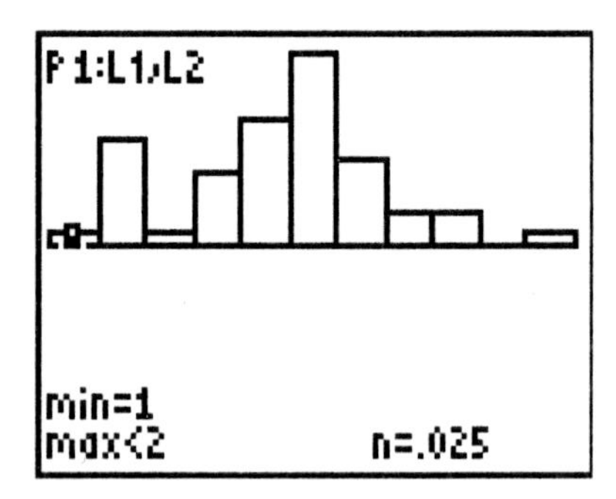

Example 5 (pg. 87) A Histogram for Continuous Data

Press **STAT** and **ENTER** to select **1:Edit**. If there is data in **L1**, highlight **L1** at the top of the first list and press **CLEAR** and **ENTER** to clear the data. You should also clear **L2**.

To create this frequency histogram you can use the raw data in Table 12 on pg. 85 or the summarized data in Table 13 also on pg. 85. For this example, we will use the summarized data. You must enter the midpoints of each class into List1 (**L1**) and the frequencies into List 2 (**L2**). To obtain the midpoints of each class, add two consecutive lower limits and divide by 2. For example, here is the calculation for the first class: (10+12)/2= 11.

To enter the midpoints in L1, you can do the calculation for the midpoints right on this screen. Simply type the calculation on the data entry line and push **ENTER**. The calculation will be automatically converted to the midpoint.

```
L1       |L2      |L3      1
-------  | ------ | ------

L1(1)=(10+12)/2
```

To set up the histogram, push **2nd** **[STAT PLOT]** and **ENTER** to select **Plot 1**. Turn ON **Plot 1**, set **Type** to **Histogram**, set **Xlist** to **L1**,. set **Freq** to **L2.**

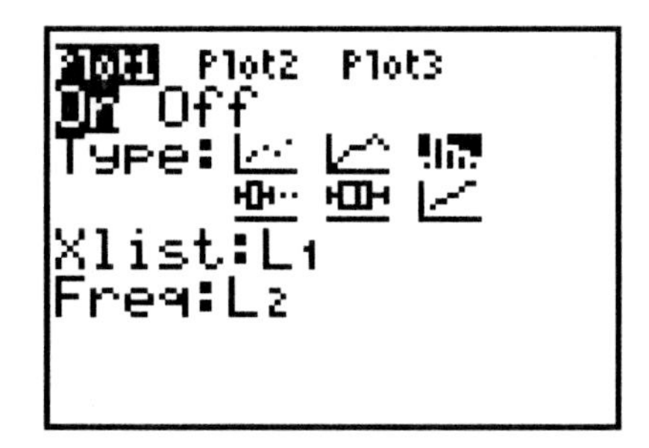

Press **ZOOM**, scroll down to **9:ZoomStat** and press **ENTER** to create a histogram. Press **Window** to adjust the Graph Window. Set **Xmin** equal to 10 (the lower limit of the first class) and **Xmax** equal to 24 (a value that would be the lower limit of an additional class at the end of the table. This extra value is needed to complete the last bar of the histogram). Set **Xscl** equal to 2, which is the class width. (Note: In many cases it is not necessary to change the values for **Ymin**, **Ymax** or **Yscl**. What you must do is to check these values and make sure that **Ymin** is a small negative value (a value between –6 and –1 would be good) and **Ymax** is slightly larger than the largest frequency value in your dataset. You never need to adjust **Yscl**.

Press GRAPH and the histogram should appear.

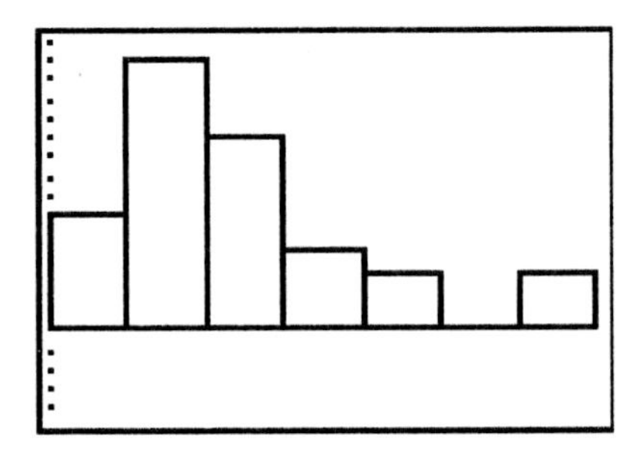

You can press TRACE and scroll through the bars of the histogram.
The minimum value of the class will appear as **Min.** **Max** is written as an inequality that states that the maximum value in the class is *less than* the given value. **n** is the number of data points in the class.

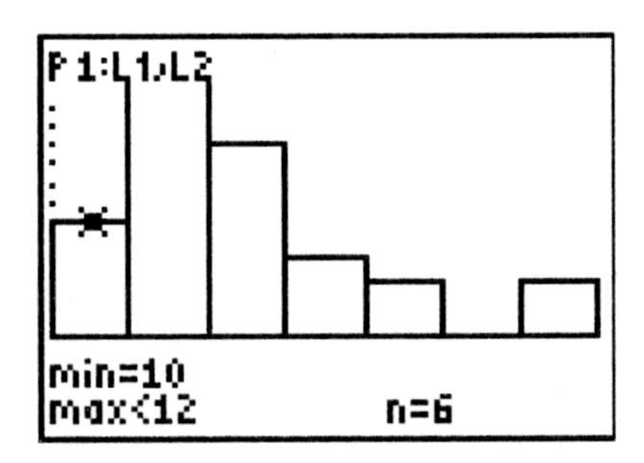

Notice, for example, with the cursor highlighting the first bar of the histogram, you will see that the first class contains values greater than or equal to 10 and less than 12 and that there are 6 data points in the this class.

To create the relative frequency histogram, replace the frequencies in L2 with relative frequencies. One way to do this is to CLEAR L2 and enter the relative frequencies from Table 13 on pg. 87. Alternately, without clearing the frequencies, move the cursor so that it highlights the label, L2, at the top of the column and press ENTER. On the data entry line, type in 2nd **L2**/40 and press ENTER.

Press Window to adjust the Graph Window. Set **Xmin** equal to 10 and **Xmax** equal to 24. Set **Xscl** equal to 2. Set **Ymin** to -.1 and **Ymax** to .40. Press GRAPH and the relative frequency histogram should appear.

You can press **TRACE** and scroll through the bars of the histogram. The minimum and maximum values of the class will appear. **n** is now the relative frequency of the class.

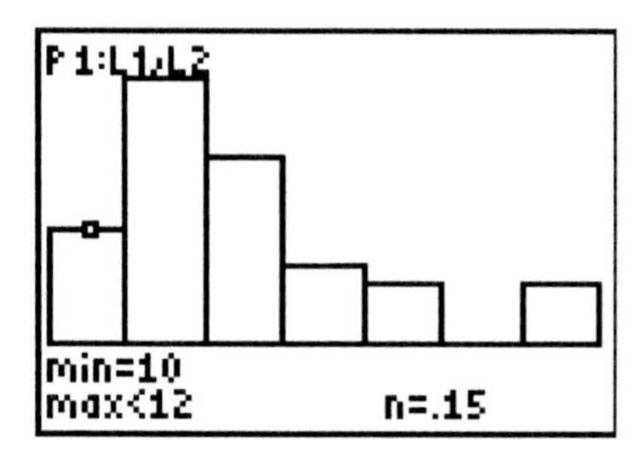

‣ Problem 34 (pg. 97)

For this example, we will construct the frequency histogram first (part c.) and then use it to find the frequencies for the frequency distribution (part a.).

Press STAT and select **1:Edit** and press ENTER. Highlight the name "**L1**" and press CLEAR and ENTER. You may also clear **L2** but you will not be using **L2** in this example. Enter the data values into **L1**.
To set up the histogram, push **2nd** [STAT PLOT and ENTER to select **Plot 1**. Turn ON **Plot 1**, set **Type** to **Histogram**, set **Xlist** to **L1.** Since you are using the raw data, you must set **Freq** to **1,** which indicates that you are entering individual data values. If the frequency is set on **L2** move the cursor so that it is flashing on **L2** and press CLEAR . The cursor is now in ALPHA mode (notice that there is an "A" flashing in the cursor). Push the ALPHA key and the cursor should return to a solid flashing square. Type in the number **1.**

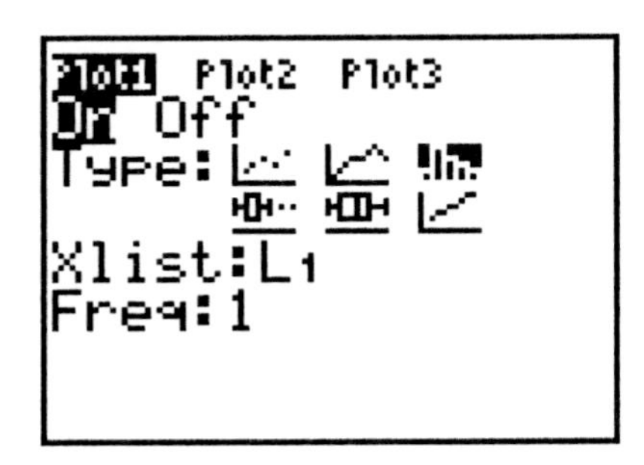

Press ZOOM , scroll down to **9:ZoomStat** and press ENTER to create a histogram. Press Window to set the Graph Window. The first value you must enter is the value for **Xmin.** This value will be the lower class limit of the first class which is **8**. The value for **Xmax** would be the lower class limit of the one extra class that would be needed to complete the last bar of the histogram. Look through the data in your textbook. Notice that the largest data point is **22.9**, therefore, the last class would be 22 – 23.9. The lower class limit of the next class would be **24**. This is the value for **Xmax**. Set **Xscl** equal to **2**, which is the class width. (Note: In many cases it is not necessary to change the values for **Ymin**, **Ymax** or **Yscl**. What you must do is to check these values and make sure that **Ymin** is a small negative value (a value between –6 and –1 would be good) and **Ymax** is slightly larger than the largest frequency value in your dataset. You never need to adjust **Yscl**.

Press GRAPH and the histogram should appear.

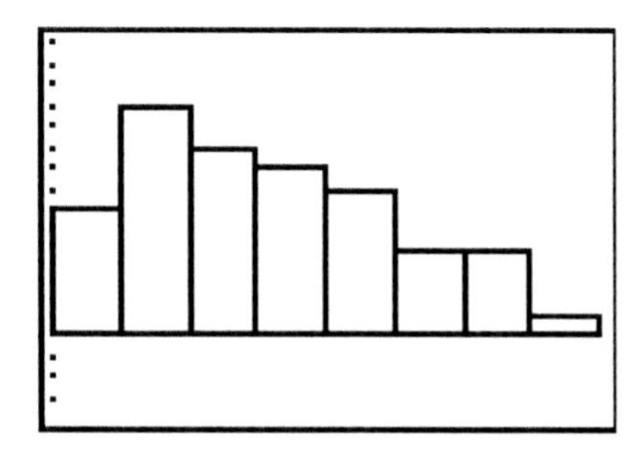

You can press TRACE and scroll through the bars of the histogram.
The minimum value of the class will appear as **Min.** **Max** is written as an inequality that states that the maximum value in the class is *less than* the given value. **n** is the number of data points in the class.

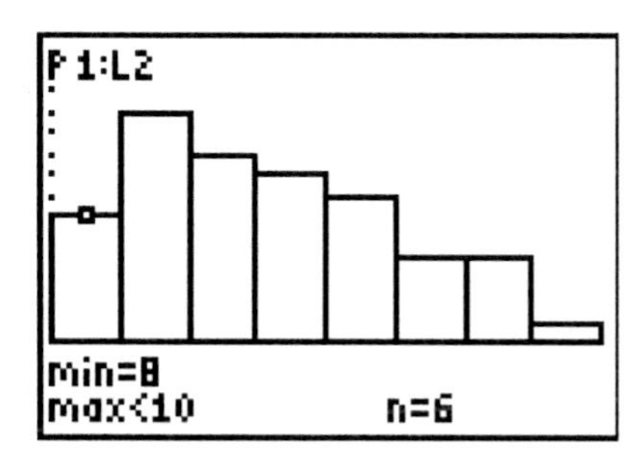

In the screen shown here, you see that the first class is 8-9.9 and there are 6 data points in this class.
Use the right arrow to scroll through the bars of the histogram and use this information to construct the frequency distribution for part (a) of this problem. Here is a starting setup for the frequency distribution table:

Class	Frequency	Relative Frequency
8-9.9	6	
10-11.9	11	

To complete the relative frequency column (for part b.), simply divide each frequency by the total frequency. For example, the relative frequency for the first class would be 6/51.

To do part (f) of the problem, press **Window** to set the Graph Window. Set **Xscl** equal to 1, set **Xmax** to 23 and set **Ymax** to 8. Press **GRAPH** and the new histogram should appear.

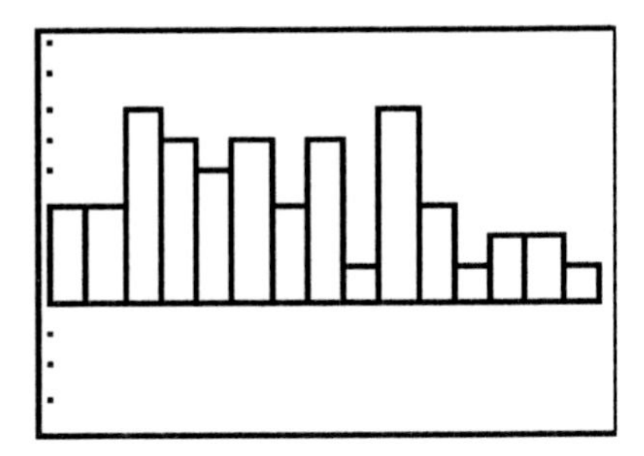

(Note: A relative frequency histogram (part d.) uses relative frequencies rather than frequencies. The actual picture on the TI-84 would be identical to the frequency histogram.)

Section 2.3

▸ Polygons (pg. 102) Constructing A Frequency Polygon

Press **STAT** and select **1:Edit** and press **ENTER**. Clear all data from **L1** and **L2**. Enter the midpoints from Table 19 on pg. 102 into **L1** and enter the frequencies into **L2**.

To set up the frequency polygon, press **2nd** **[STAT PLOT]**. Press **ENTER** to select **Plot 1**. Highlight **On** and press **ENTER**. Set **Type** to the frequency polygon which is the second selection and press **ENTER**. Set **Xlist** to **L1** and **Freq** to **L2**. Next, there are three different types of **Marks** that you can select for the graph. The first choice, a small square, is the best one to use.

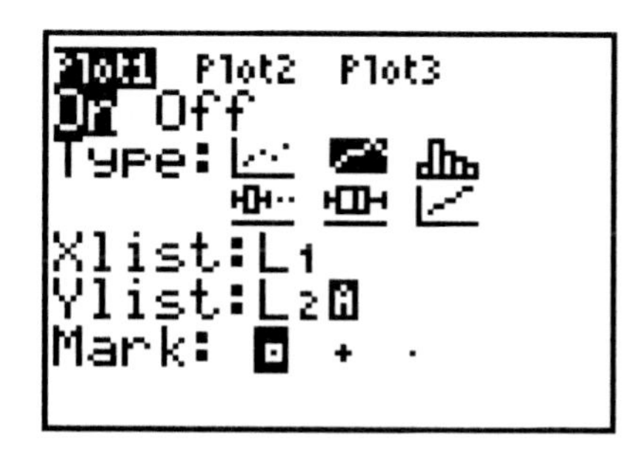

Press **ZOOM** and scroll down to **9:ZoomStat** and press **ENTER** or simply press **9** and **ZoomStat** will automatically be selected.

You can press **TRACE** and scroll through the points in the polygon. For example, the third data point represents class 3 which has a midpoint of 15 and a frequency of 10.

◀

Ogives (pg. 104) Constructing A Relative Frequency Ogive

Press **STAT** and select **1:Edit** and press **ENTER**. Clear all data from **L1** and **L2**. We will be using the data in Table 20 on pg. 104. An ogive uses the upper class limits for the x-coordinates. An additional upper class limit for the class that would precede the first class is included. For this example that value would be 9.99, and that is the first data point entered in **L1**. Follow this data point with the upper class limits from Table 20. The y-coordinates are the cumulative relative frequencies and they are entered in **L2**. The y-coordinate that corresponds to the additional upper class limit of 9.99 is 0. This additional data point (9.99, 0) is included so that the first line segment in the ogive starts on the x-axis.

L1	L2	L3 2
9.99	0	------
11.99	.15	
13.99	.5	
15.99	.7	
17.99	.85	
19.99	.925	
21.99	.925	

L2(1)=0

To set up the ogive, press **2nd** **[STAT PLOT]**. Press **ENTER** to select **Plot 1**. Highlight **On** and press **ENTER**. Set **Type** to the frequency polygon which is the second selection and press **ENTER**. Set **Xlist** to **L1** and **Freq** to **L2**.

Press **ZOOM** and scroll down to **9:ZoomStat** and press **ENTER** or simply press **9** and **ZoomStat** will automatically be selected.

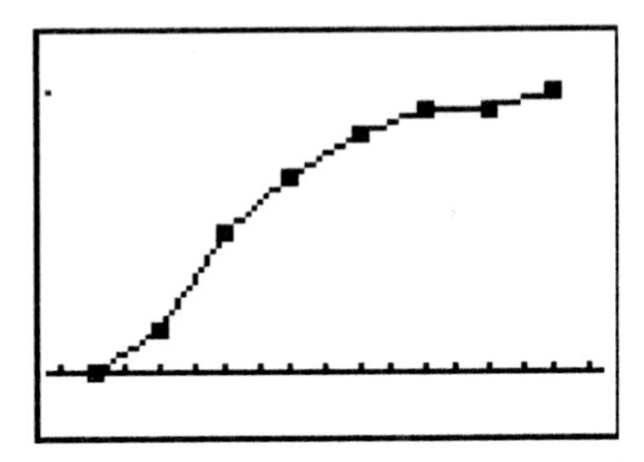

You can press **TRACE** and scroll through the points in the ogive.

Example 1 (pg. 105) A Time Series Plot

Press **STAT** and select **1:Edit**. Clear **L1** and **L2**. Notice that the dates in Table 21 on pg. 105 are "1/06" through "12/07." Rather than entering these actual dates into **L1**, you can simply number the months from "1" to "24" and enter these numbers into **L1**. Enter the closing prices into **L2**.

L1	L2	L3 2
1	18.57	------
2	20.24	
3	21.67	
4	20.95	
5	19.68	
6	19.53	
7	17.88	

L2(1)=18.57

To construct the time series chart, press **2nd** **[STAT PLOT]** and select **1:Plot 1** and **ENTER**. Turn ON **Plot 1**. Set the **Type** to **frequency polygon**. For **Xlist** select **L1** and for **Ylist** select **L2**. Press **ZOOM** and scroll down to **9:ZoomStat** and press **ENTER** or simply press **9** and **ZoomStat** will automatically be selected.

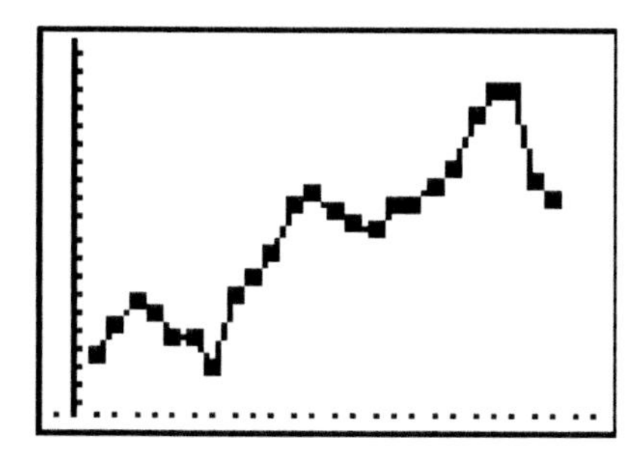

You may want to adjust the Window so that the Y-axis has a lower limit of 0. Simply press **Window** and change **Ymin** to 0. Press **GRAPH** and the time series plot should appear.

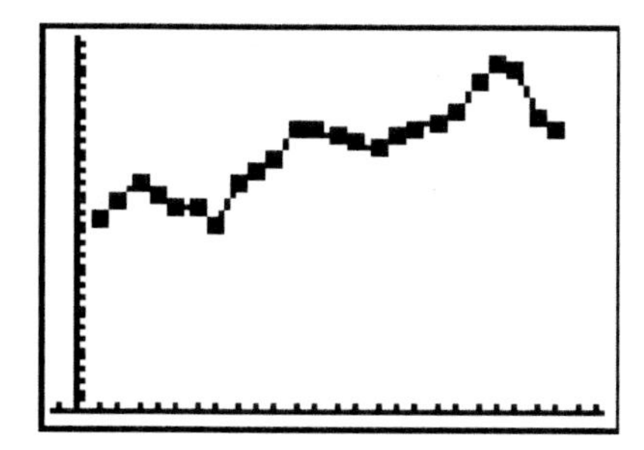

Use TRACE to scroll through the data values for each year. Notice for example, the closing price for the 4th month (which is April 2006) is 20.95.

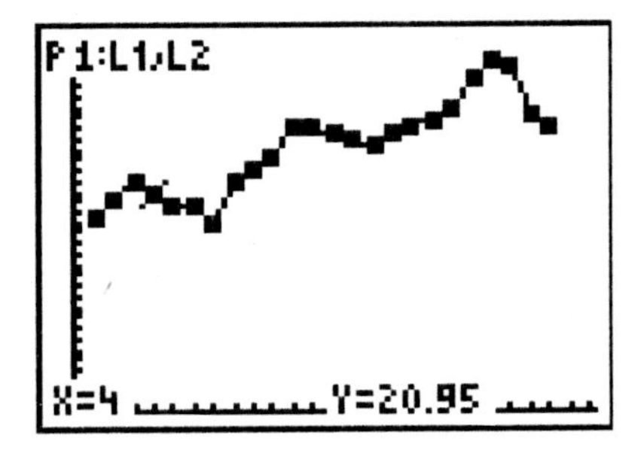

‣ Problem 20 (pg. 110)

For this exercise, use the dataset in Problem34 on pg. 97 and the frequency distribution table from pg. 23 of this manual.

Class	Midpoint	Frequency	Relative Frequency	Cumulative Frequency	Cumulative Relative Frequency
8-9.9	9.0				
10-11.9	11.0				

a.) Enter the remaining classes, the midpoints of each class and the frequencies and relative frequencies for each class. To complete the cumulative frequency column, simply accumulate the frequencies. For example, the cumulative frequency for the third class would be the sum of the frequencies for the 1st, 2nd and 3rd classes.

b.) To complete the cumulative relative frequency column, accumulate the relative frequencies.

c.) Press STAT and select **1:Edit**. Clear the lists and enter the midpoints into **L1** and the frequencies into **L2.** Press **2nd** [STAT PLOT] and select **Plot 1** and press ENTER. Set the **Type** to **frequency polygon**. Set **Xlist** to **L1** and Freq to **L2. Press** ZOOM and scroll down to **9:ZoomStat** and press ENTER or simply press 9 to automatically select **ZoomStat.**

d.) Press STAT and select **1:Edit** and press ENTER. Clear all data from **L1** and **L2.** Enter the upper class limits into **L1** and enter the cumulative frequencies into **L2**. Press **2nd** [STAT PLOT] and select **Plot 1** and press ENTER. Set the **Type** to **frequency polygon**. Set **Xlist** to **L1** and Freq to **L2. Press** ZOOM and scroll down to **9:ZoomStat** and press ENTER or simply press 9 and **ZoomStat** will automatically be selected.

e.) A relative frequency ogive is constructed using the upper class limits and the cumulative relative frequencies (rather than the cumulative frequencies). The actual picture on the TI-84 is identical to the frequency ogive that you constructed in part (d).

TI-89 Instructions:

These instructions are designed to give you an overview of the statistical graphing capabilities of the TI-89.

From the **Home** screen, select the **Stat/List** icon. The screen that follows allows you to store the data in specific folders. It is easy to always use the **Main** folder so simply press ENTER. The next screen displays Lists 1 through 4. If any Lists contain data and you want to delete that data, press **F3**, select **1:Names** and scroll down to each **List** that you would like to delete. While the list is highlighted, press **F4** to select it. Once you have selected all the lists that you want to delete, press **F1** and select **1:Delete** and press ENTER. Press ESC to return to the lists. Notice that the lists that you just deleted are no longer displayed. Press **F1** and select **3:Setup Editor** and press ENTER. Lists 1 through 4 are again displayed. Now you are ready to enter the data. If you have a single set of data values, enter them into **List1**. If you have a table of data values along with their frequencies, enter the data into **List1** and enter the frequencies into **List2**. If you have bivariate data enter the x-values into **List1** and the y-values into **List2**.

Press **F2** and select **1:Plot Setup** and press **F1** to **Define your plot**. On the Plot Setup screen, use the right arrow to display the **Plot Type** options. Select the type that you want and press ENTER. Suppose that you choose the Histogram. Move to the first entry box to define the x-variable (list1) and press 2nd VAR-LINK. Select the column name that contains the x-variable and press ENTER. Move to the line: 'Hist.Bucket Width' and enter the class width for your histogram. Move to the line: "Use Freq and Categories.' If you are creating the histogram from the raw data, select **No**. If you are creating the histogram from a table with x-values and their frequencies, select **Yes**. If you select **Yes**, move to the next line, press 2nd VAR-LINK and select the column name that contains the frequencies (list2) and click ENTER until you return to the previous screen. Press **F5** for **Zoom Data**. You can adjust the graph window by pressing ◆ and **F2.** Set **Xmin** to the smallest x-value. Set **Xmax** to a value slightly larger than the largest data value (this is necessary to complete the last bar of the histogram). Set **Xscl** to the class width. Set **Ymin** to a small negative value (a value between –6 and –1 would be good). **Ymax** should be set to a value that is slightly larger than the largest frequency value in your dataset. Press ◆ and **F3** to graph.

Notice the **TRACE** option which is F3 at the top of the graph. If you press it, a flashing cursor will appear on the screen. In the case of the histogram, the cursor appears at the top of the 1st bar of the histogram. Notice the information at the bottom of the screen. **Min** and **Max** define the first bar of the histogram. "n= " tells us the number of data values in that particular bar of the histogram. You can use the right arrow to move through each of the bars.

TI-*n*Spire Instructions:

These instructions are designed to give you an overview of the single variable graphing capabilities of the TI-*n*Spire.

Press (home) and select **6:New Document**. (Note: If you currently have a document open, the next screen will ask if you want to save the document. Press (tab) to select **No**. Press (enter).) Select **3:Add Lists & Spreadsheet. For a single set of data values:** move to the top of column **A** and in the box next to '**A**' type in a name. Move to **Line 1** in column **A** and begin entering your data values. **For bivariate data**: move to the top of column **A** and in the box next to '**A**' type in a name. Move to **Line 1** in column **A** and begin entering the x-values. Move to the top of column **B** and in the box next to '**B**' type in a name. Move to **Line 1** in column **B** and begin entering the y-values. (Note: If you have **a table of data values along with their frequencies,** the TI-*n*Spire does not at this time have the graphing capability to graph this data.)

To construct a graph of your data, press (home) and select **5: Data & Statistics**. A new page will open with a scatter of points. Press and hold the down arrow to move the cursor to the middle of the x-axis at the bottom of the screen. When the box '**Click to add variable**' appears, press (enter). Highlight the Column **A** name and press (enter). A Dotplot of the data in Column A will be displayed. To adjust the Window, press (menu), select **4:Window/Zoom** and select from the various Window options. To look at other graphs of the data in column A, press (menu) and select **1:Plot Type** and select the desired plot.

You can use the arrow keys to move through a graph and highlight key characteristics. For example, in a histogram, move to the top of one of the bars until the arrow cursor changes to (hand). Press and hold (click). The minimum and maximum values for that particular bar of the histogram will appear along with the number of data values in that particular bar. Press and release (click) to return to the (hand) cursor and use the arrow keys to move to another location. Press and hold (click) to display information on that location.

For bivariate data, once you have set up a graph for the data in Column A, press (menu), select **2:Plot Properties** and select **6:Add Y Variable.** A list of column names will appear. Highlight the name for column **B** and press (enter). A scatterplot of the data will be displayed. Use the arrow keys to move through the data points. Once you move to a specific data point, the arrow cursor changes to (hand). Press and hold (click). The x and y coordinates for that data point will be displayed. Press and release (click) to return to the (hand) cursor and use the arrow keys to move to another data point.

Numerically Summarizing Data

Section 3.1

▸ Example 1 (pg. 130) A Population Mean and a Sample Mean

The TI-84 has one method for calculating the mean of a dataset. This method is used for a population mean, μ, and a sample mean, $\bar{x}$. The symbol that the calculator uses for the mean of a dataset is always $\bar{x}$. If you are calculating the mean of a population, then the value for $\bar{x}$ that you obtain from the calculator is actually the value for μ.

(a). Press **STAT** and select **1:Edit**. Clear **L1** and enter the scores for the ten students into **L1**. Press **STAT** again and highlight **CALC** to view the Calc Menu.

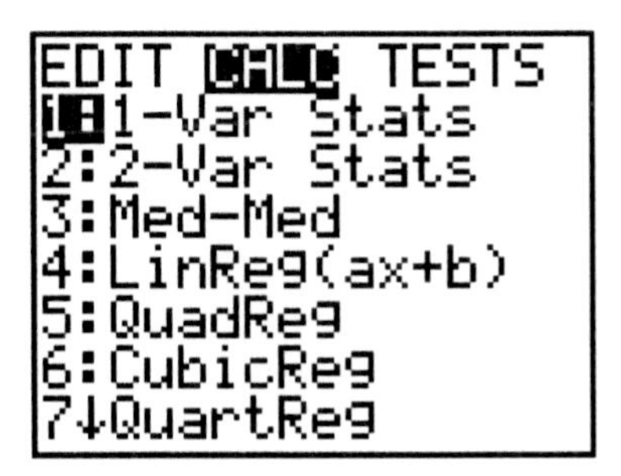

Select **1:1-Var Stats.** On this line, enter the name of the column that contains the data. Since you have stored the data in **L1,** simply enter **2nd** **[L1]** **ENTER** and the first page of the one variable statistics will appear. (Note: If you did not enter a column name, the default column, which is **L1,** would be automatically selected.)

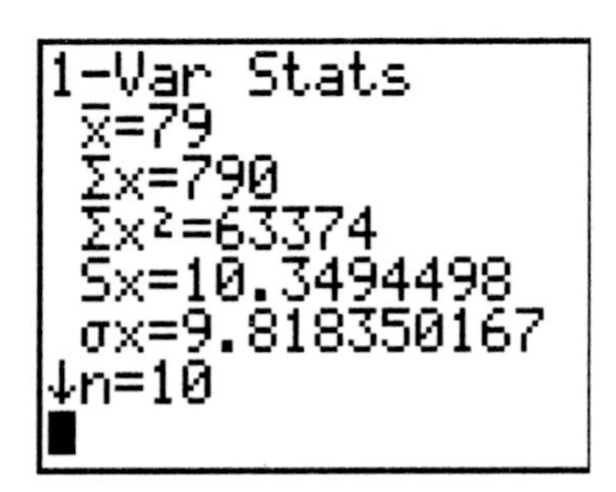

The first item is the mean of the dataset. The correct symbol for the mean of a population is μ. For this dataset, $\mu = 79$.

(b). To use a seed of '54', press **54** **STO** **MATH** and select **PRB**. Press **ENTER** to select **1:Rand** and press **ENTER**. (Note: Selecting a seed is an optional step which can be omitted when generating random data.)

To generate a random sample of 4 students from the 10 students, press **MATH** and select **PRB**. Select **5:RandInt** by pressing **'5'** or moving the cursor to **5:RandInt** and pressing **ENTER**. Enter a starting value of **1**, an ending value of **10** and a sample size of **8**. (Recall: The TI-84 samples with replacement. This method may result in duplicates in your sample. Selecting a few more values than you need is a good practice so that you will obtain the necessary number of distinct values for your sample.)

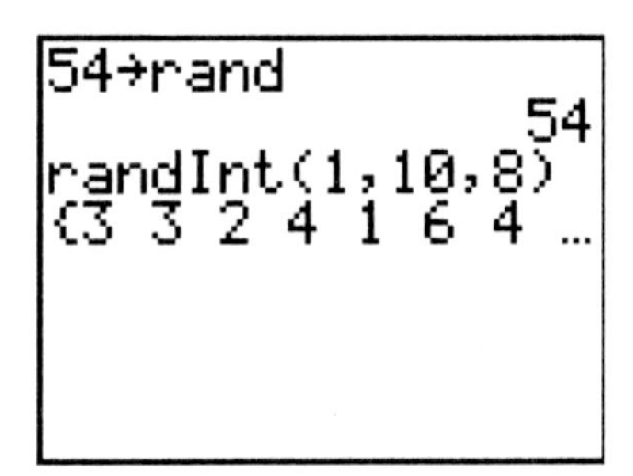

The random sample of 4 students is the first 4 distinct numbers: 3, 2, 4, and 1. (Note: If you had not selected '54' as the seed, you would have obtained a different set of random numbers.)

(c.) Press **STAT** and select **1:Edit**. Clear **L2** and enter the data values for the 4 randomly selected students into **L2**. Press **STAT** again and highlight **CALC** to view the Calc Menu. Select **1:1-Var Stats.** On this line, enter the name of the column that contains the data. Since you have stored the data in **L2,** simply enter **2nd** **[L2]** **ENTER** and the first page of the one variable statistics will appear.

```
1-Var Stats
 x̄=80
 Σx=320
 Σx²=25794
 Sx=8.041558721
 σx=6.964194139
↓n=4
```

The sample mean, $\bar{x}$, is 80.

▸ Example 2 (pg. 132) The Median of a Dataset

Press **STAT** and select **1:Edit**. Clear **L1** and enter the data from Table 2 on pg. 132 into **L1**. Press **STAT** again and highlight **CALC** to view the Calc Menu. Select **1:1-Var Stats** and press **2nd** **[L1]** **ENTER**. Notice the small arrow in the bottom left corner of the screen. This indicates that more information follows this first page. Use the down arrow to scroll through this information. The third item you see on the second page is the median, Med = 217.

```
1-Var Stats
↑Sx=31.88172587
 σx=30.05837941
 n=9
 minX=179
 Q1=203.5
↓Med=217
```

(Note: It is not necessary to put the data in ascending order when calculating a median using the TI-84.)

◂

▸ Problem 30 (pg. 140)

Press **STAT** and select **1:Edit**. Clear **L1** and enter the data into **L1**. Press **STAT** again and highlight **CALC** to view the Calc Menu. Select **1:1-Var Stats** and press **2nd** **[L1]** **ENTER**.

(a.) The first item is the sample mean, 104.1 seconds. To find the median, scroll down to the next page of output. The median is 104 seconds.

(b.) Before graphing the histogram, make sure that there are no functions in the Y-registers. To do this, press the **Y=** key. If there are any functions stored in any of the Y-values, simply move the cursor to the line that contains a function and press **CLEAR**. Now you are ready to graph the histogram. Press **2nd** **[STAT PLOT]** (located above the **Y=** key).

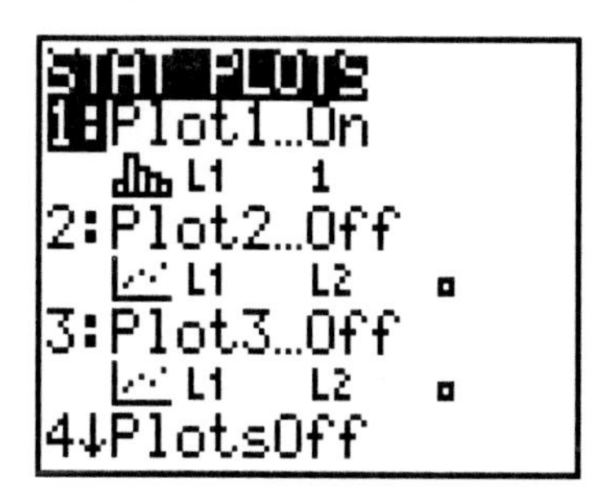

Select Plot1 by pressing **ENTER**. Position the cursor on **ON** and press **ENTER** to select it. On the next line, move your cursor to the symbol for histogram (3rd item in the 1st line of **Type**) and press **ENTER**. The next line is **Xlist**. Use the down arrow to move to this line. On this line, you tell the calculator where the data values are stored. In most graphing situations, the data are entered into **L1** so **L1** is the default option. Notice that the cursor is flashing on **L1**. Push **ENTER** to select **L1**. The last line is the frequency line. On this line, **1** is the default. The cursor should be flashing on **1** which indicates that you are entering the individual data values.

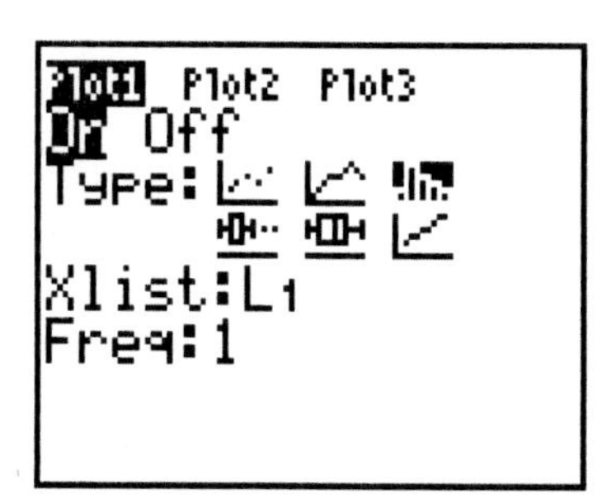

To view a histogram of the data, press **ZOOM** and scroll down to **9:ZoomStat** and press **ENTER** or simply press **9** and **ZoomStat** will automatically be selected.

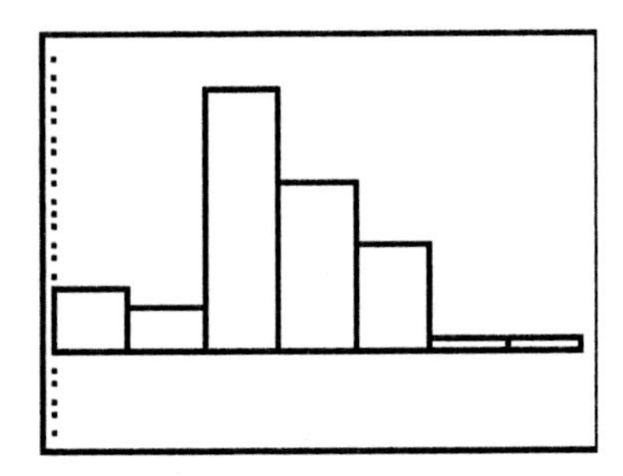

◀

Section 3.2

Example 6 (pg.149) The Variance and Standard Deviation

Press **STAT** and select **1:Edit**. Clear **L1** and enter the population data from Table 8 on pg. 144 into **L1**. Press **STAT** and highlight **CALC** to display the Calc Menu. Select **1: 1-Var Stats** and press **2nd** **[L1]** **ENTER**.

```
1-Var Stats
 x̄=79
 Σx=790
 Σx²=63374
 Sx=10.3494498
 σx=9.818350167
↓n=10
```

The population standard deviation is σx, 9.818350167.
To calculate the population variance, $(\sigma x)^2$, type in the value of the standard deviation at the bottom of the screen and press the x^2 key and **ENTER**.

```
 Σx=790
 Σx²=63374
 Sx=10.3494498
 σx=9.818350167
↓n=10
9.818350167²
                96.4
```

The population variance is 96.4.

Next, obtain the sample standard deviation and sample variance for the sample of four students. Press **STAT** and select **1:Edit**. Clear **L2** and enter the sample data points (90, 77, 71 and 82) into **L2**. Press **STAT** and highlight **CALC** to display the Calc Menu. Select **1: 1-Var Stats** and press **2nd** **[L2]** **ENTER**.

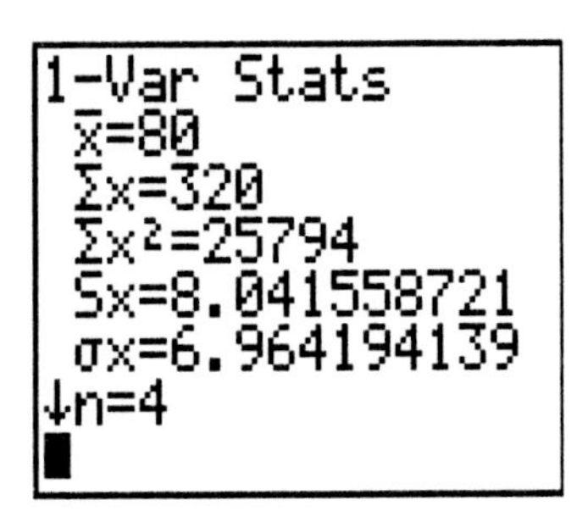

The sample standard deviation is Sx, 8.041558721.
To calculate the sample variance, $(\text{Sx})^2$, type in the value of the standard deviation at the bottom of the screen and press the x^2 key and **ENTER**.

Problem 25 (pg. 155)

(a.) Press **STAT** and select **1:Edit**. Clear **L1** and enter the data into **L1**. Press **STAT** and highlight **CALC** to display the Calc Menu. Select **1: 1-Var Stats** and press **2nd** **[L1]** **ENTER**. The population standard deviation is σx, 7.67.

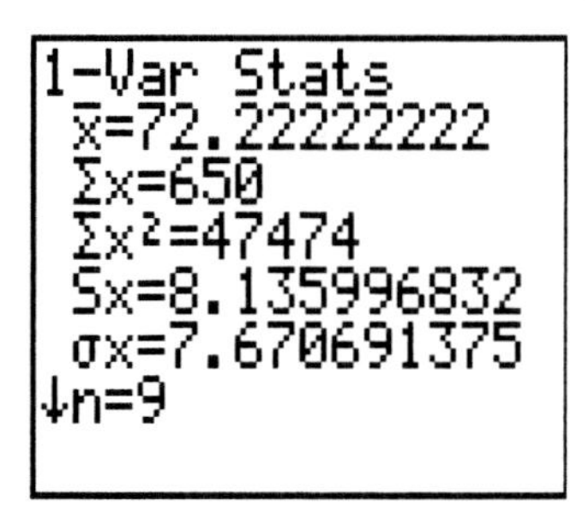

(b.) To generate a random sample of size 3, first number the students from 1 to 9. The TI-84 samples with replacement, so you will need to generate a sample that is larger than 3. Suppose you take a sample of size 5. Press **MATH** and **PRB**. Select **5:RandInt** and enter **1, 9, 5**. Here is one possible outcome.

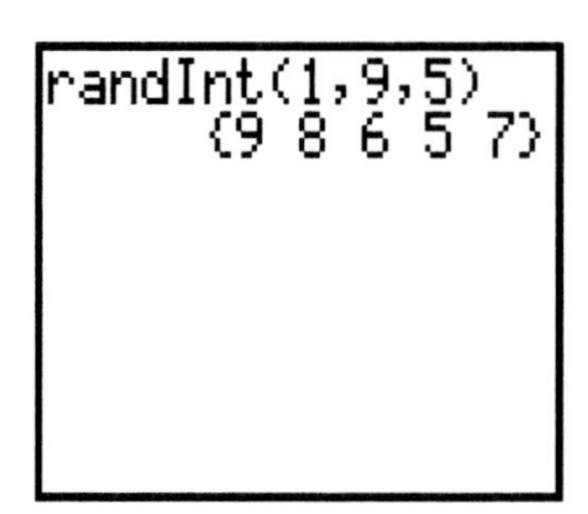

In this outcome we have selected 5 students: Student Numbers: 9,8,6,5 and 7. Since we need only need 3 students, simply select the first three: 9, 8, and 6. (Note: Random samples vary so your results will be different.)

Enter the pulse rates of the 3 students into L2. Press **STAT** and highlight **CALC** to display the Calc Menu. Select **1: 1-Var Stats** and press **2nd** **[L2]** **ENTER**. The sample standard deviation is Sx, 6.0277.

Repeat this process to generate two more random samples of size 3.

▸ Problem 42 (pg. 157)

(a.) Press **STAT** and select **1:Edit**. Clear **L1** and enter the salary data into **L1**. Press **STAT** and highlight **CALC** to display the Calc Menu. Select **1: 1-Var Stats** and press **2nd [L1]** **ENTER**. The population standard deviation, σx, is 12.6. The population variance, $(\sigma x)^2$, is 160. The range (75-30) is 45.

(b.) Press **STAT** and select **1:Edit**. Clear **L2.** Move the cursor to the top of **L2** again and press **ENTER**. The cursor should now be flashing on the bottom line of the screen. On this line, type in **2nd [L1] + 2.5.**

L1	L2	L3 2
30	------	------
30		
45		
50		
50		
50		
55		

L2 =L1+2.5

Press **ENTER**. Each value in **L2** should be 2.5 thousand dollars higher than the corresponding value in **L1**. Press **STAT** and highlight **CALC** to display the Calc Menu. Select **1: 1-Var Stats** and press **2nd [L2]** **ENTER**. The population standard deviation, σx, is 12.6, the same as the standard deviation for the original data. The range and variance also remain unchanged.

(c.) Press **STAT** and select **1:Edit**. Clear **L3.** Move the cursor to the top of **L3** again and press **ENTER**. The cursor should now be flashing on the bottom line of the screen. On this line, type in **2nd [L1] *1.05.** Press **STAT** and highlight **CALC** to display the Calc Menu. Select **1: 1-Var Stats** and press **2nd** [L3] **ENTER**. The population standard deviation, σx, is 13.3, which is 1.05 times the value of the standard deviation for the original data. The variance and range are also 1.05 times the variance and range of the original data.

(d.) Press **STAT** and select **1:Edit**. In **L1**, move the cursor so that it is flashing on 75 (Benjamin's salary). Type in **100**. Press **STAT** and highlight **CALC** to display the Calc Menu. Select **1: 1-Var Stats** and press **2nd [L1]** **ENTER**. The population standard deviation, σx, has increased to 18.5. The population standard variance, $(\sigma x)^2$, has increased to 341.3. The range has increased to 70.

◂

Section 3.3

Example 2 (pg. 162) The Weighted Mean

Press **STAT** and select **1:Edit**. Clear **L1** and **L2.** Enter the point values for each letter grade that Marissa earned into **L1**. Enter the corresponding credits earned into **L2**.

L1	L2	L3 2
4	4	------
3	3	
4	3	
2	5	
4	1	

L2(6) =

Press **STAT** and highlight **CALC** to display the Calc Menu. Select **1: 1-Var Stats** and press **2nd** **[L1]** , **2nd** **[L2]**. Press **ENTER**. (Note: You must place the comma between **L1** and **L2**).

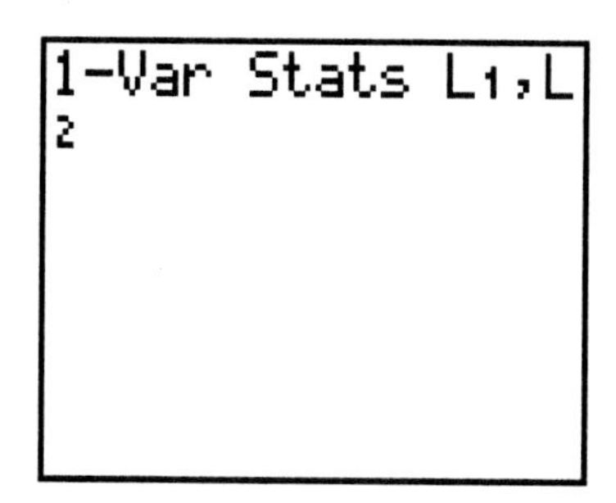


```
1-Var Stats
 x̄=3.1875
 Σx=51
 Σx²=175
 Sx=.910585892
 σx=.8816709987
↓n=16
```

Her GPA (weighted average) is 3.1875.

Example 4 (pg. 164) The Mean and Standard Deviation of a Frequency Distribution

Press **STAT** and select **1:Edit**. Clear **L1** and **L2.** Refer to Table 13 on pg. 161. Calculate the midpoints of each class by adding consecutive lower class limits and dividing by 2. You can do the calculations for the midpoints directly on this screen. For the first class, type in **(10+12)/2** and press **ENTER**.

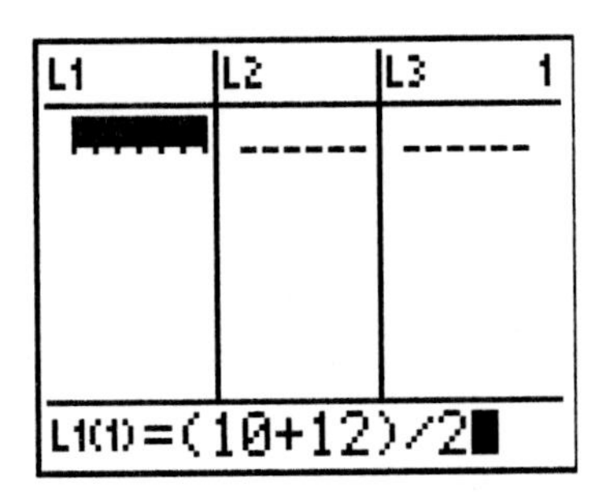

The value of the midpoint, 11, will appear as the first entry in **L1**. Continue this process to obtain the midpoints for each of the classes. Enter the frequencies into **L2**. Press **STAT** and highlight **CALC** to display the Calc Menu. Select **1: 1-Var Stats** and press **2nd** **[L1]** **,** **2nd** **[L2]**. Press **ENTER**. (Note: You must place the comma between **L1** and **L2**).

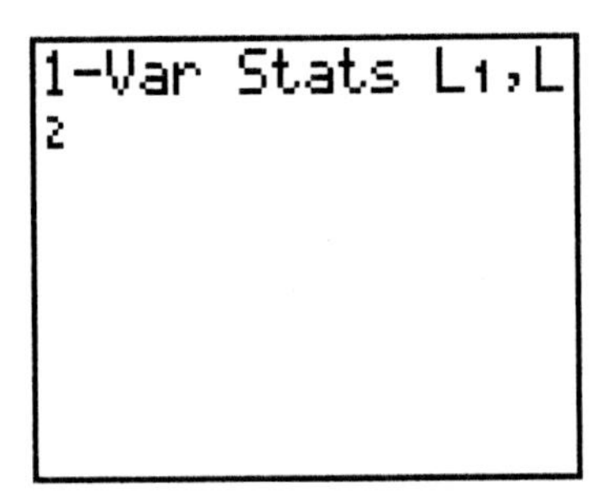

Using **L1** and **L2** in the **1:1-Var Stats** calculation is necessary when approximating a mean from a frequency distribution. The calculator uses the data in **L1** and the associated frequencies in **L2** to approximate the sample statistics for the dataset. In this example, the approximate mean is 14.8 percent and the approximate standard deviation is 3.23 percent.

▸ Problem 7 (pg. 165)

(a.) Press **STAT** and select **1:Edit**. Clear **L1** and **L2**. Enter the midpoints for each of the age ranges into **L1** and the frequencies ('number of multiple births') into **L2**. Press **STAT**, highlight **CALC**, select **1:1-Var Stats**, and press **2nd** **[L1]** **,** **2nd** **[L2]** **ENTER**.

L1	L2	L3 2
17.5	84	------
22.5	431	
27.5	1753	
32.5	2752	
37.5	1785	
42.5	378	
47.5	80	

L2(1)=84

The population statistics will appear on the screen.

```
1-Var Stats
 x̄=32.46562156
 Σx=236090
 Σx²=7874650
 Sx=5.372154829
 σx=5.371785443
↓n=7272
```

The approximate value of the population mean is 32.47 and the approximate value of the population standard deviation (σx) is 5.37.

(b.) To set up the histogram, push **2nd** **[STAT PLOT]** and **ENTER** to select **Plot 1**. Turn ON **Plot 1**, set **Type** to **Histogram**, set **Xlist** to **L1,** set **Freq** to **L2.**

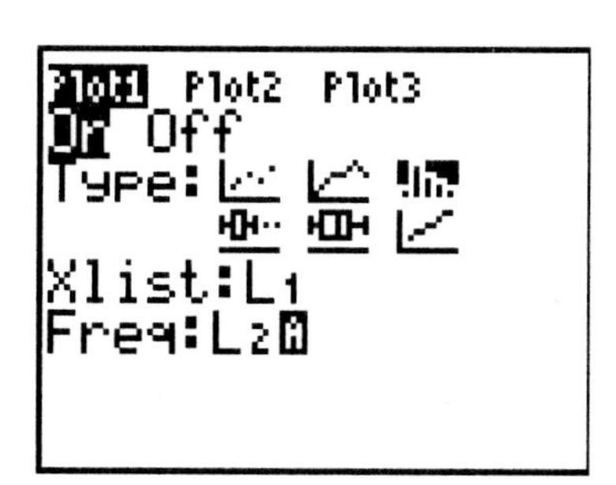

Press **Window** to adjust the Graph Window. Set **Xmin** equal to 17.5 (the midpoint of the first class) and **Xmax** equal to 57.5 (a value that would be the midpoint of an additional class at the end of the table. This extra value is needed to complete the last bar of the histogram). Set **Xscl** equal to 5, which is the class width. Set **Ymin = -20** and **Ymax = 2800**. You do not need to change **Yscl** or **Xres**. Press **GRAPH** and the histogram should appear.

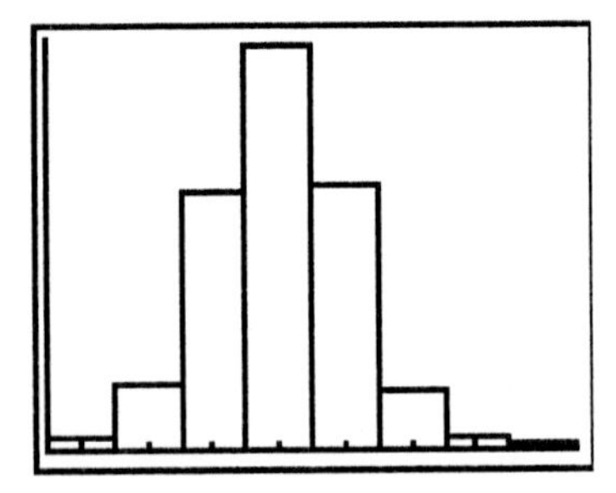

It may look as if there are only seven bars in the histogram. There are actually 8 bars. The last bar has such a small frequency compared to the other bars that it is extremely small in the graph. Notice that the histogram is bell-shaped.

(c.) The Empirical Rule states that 95% of the data falls in the interval $(\mu \pm 2\sigma)$. To calculate the upper and lower limits of this interval, press **CLEAR** a few times until you get a blank screen. Enter **32.5 - 2*5.4** to get the lower limit. Press **2nd** **ENTER** and the calculation will appear again on the screen with the cursor flashing. Move the cursor so that it is positioned on the '-'sign and type in a '+' sign and press **ENTER** to get the upper limit.

◀

Section 3.4

▸ Example 4 (pg. 170) Quartiles

Press **STAT** and select **1:Edit**. Clear **L1.** Enter the collision cover claims data from Table 16 on pg. 169 into **L1**. Press **STAT**, highlight **CALC**, select **1:1-Var Stats**, and press **2nd [L1] ENTER**. Scroll down to the 2nd screen to find the quartiles. **Q1** is the first quartile, **Med** is the 2nd quartile (or median) and **Q3** is the third quartile.

```
1-Var Stats
↑n=18
 minX=180
 Q1=735
 Med=1805
 Q3=4668
 maxX=21147
```

◂

Problem 21 (pg. 174)

Press **STAT** and select **1:Edit**. Clear **L1** and then enter the data. Press **STAT** and highlight **CALC**. Select **1:1-Var Stats** and press 2nd [**L1**] **ENTER**. The sample mean and sample standard deviation appear on the first screen. Scroll down to the 2nd screen to find the quartiles. **Q1** is the first quartile, **Med** is the 2nd quartile (or median) and **Q3** is the third quartile.

(a.) To calculate the Z-score for the data point 0.97 inches, do the following calculation: (0.97-sample mean)/sample standard deviation.

(c.) The interquartile range (**IQR**) is **Q3-Q1**.

(d.) The lower fence is **Q1-1.5*IQR.** The upper fence is **Q3+1.5*IQR.**

◀

Section 3.5

Example 2 (pg. 177) The Five Number Summary

Press **STAT** and select **1:Edit**. Clear **L1** and enter the data from Table 17 on pg. 176 into **L1.** Press **STAT** and highlight **CALC** to display the Calc Menu. Select **1: 1-Var Stats** and press 2nd [L1] **ENTER**. Scroll down to the 2nd screen to obtain the five values: **minX, Q1, med, Q3 and maxX.**

```
1-Var Stats
↑n=21
 minX=19.95
 Q1=26.055
 Med=30.95
 Q3=37.24
 maxX=64.63
```

▸ Example 3 (pg. 177) A Boxplot

This is a continuation of Example 2. In Example 2 we entered the data from Table 17 on pg. 176 into **L1**.

Press **2nd** [STAT PLOT]. Select **1:Plot 1** and press ENTER. Turn On **Plot 1**. Move to the **Type** options. Using the right arrow (you can not use the down arrow to drop to the second line), scroll through the **Type** options and choose the first boxplot which is the first entry in row 2 of the **TYPE** options. Press ENTER. Move to **Xlist** and type in **L1**. Press ENTER and move to **Freq**. Set **Freq** to **1.** If **Freq** is set on **L2**, press CLEAR, and press ALPHA to return the cursor to a flashing solid rectangle and type in **1**. Press ZOOM and 9 to select **ZoomStat.** The Boxplot will appear on your screen.

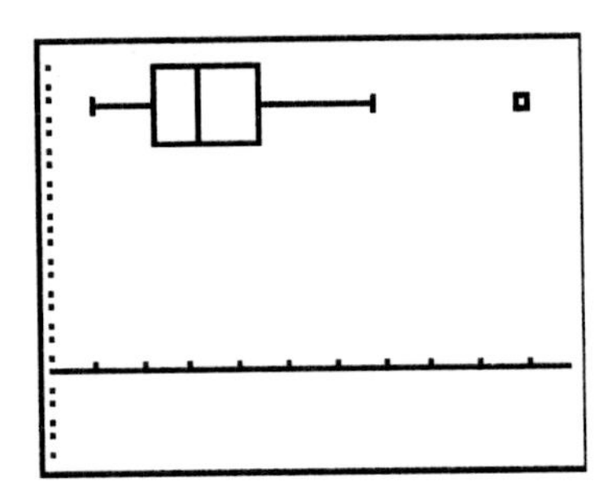

If you press TRACE and use the left and right arrow keys, you can display the following information: the smallest data point which is 19.95; Q1 (26.055); the median (30.95); Q3 (37.24); the largest data point which falls inside the upper fence which is 49.17 and the largest data point in the dataset, which is 64.63. This largest value is also an outlier because it lies outside the upper fence. Notice that this value is designated by a small box at the extreme right side of the diagram. (Note: The boxplot does not display the lower and upper fences.)

◂

TI-89 Instructions

These instructions are designed to give you an overview of the univariate statistical calculations on the TI-89.

Select the **Stat/List** icon. The screen that follows allows you to store the data in specific folders. It is easy to always use the **Main** folder so simply press ENTER. The next screen displays Lists 1 through 4. If any Lists contain data and you want to delete that data, press **F3**, select **1:Names** and scroll down to each List that you would like to delete. While the list is highlighted, press **F4** to select it. Once you have selected all the lists that you want to delete, press **F1** and select **1:Delete** and press ENTER. Press ESC to return to the lists. Notice that the lists that you just deleted are no longer displayed. Press **F1** and select **3:Setup Editor** and press ENTER. Lists 1 through 4 are again displayed. Now you are ready to enter the data. If you have a single set of data values, enter them into **List1**. If you have a table with data values along with their frequencies, enter the data into **List1** and enter the frequencies into **List2**.

Press F4, choose **1: OneVar** and press ENTER. Move to the first entry box to define the x-variable and press 2nd VAR-LINK. Select the column name that contains the x-variable (list1) and press ENTER. Move to the next line, **Freq.** If you have a single set of data values that you entered into List1, then set **Freq.** to **1**. If you have a table with data values in List1 and frequencies in List2, set **Freq** to **List2** by pressing 2nd VAR-LINK and selecting **List2**. Click ENTER until the summary statistics are displayed. Use the down arrow to view additional output that does not appear on the first output screen.

TI-*n*spire Instructions:

These instructions are designed to give you an overview of the univariate statistical calculations on the **TI-*n*spire** handhelds.

Press (home) and select **6:New Document**. (Note: If you currently have a document open, the next screen will ask if you want to save the document. Press (tab) to select **No**. Press (enter).) Select **3:Add Lists & Spreadsheet.** If you have a single set of data values, enter them into **column A.** If you have a table with data values along with their frequencies, enter the data into **column A** and enter the frequencies into **Column B**. Press (menu), select **4:Statistics** and select **1: Stat Calculations.** Select **1:One-Variable Statistics**. On the next screen, set **Num of Lists** to **1**. Press (tab) to select **OK** and press (enter). On the next screen, in the first entry box (**X list**) type **a** (for Column A). Press (tab) to scroll to the next line (**Frequency**) and type **1** (if you have a single set of data values stored in Column A) or **b** (if you have the frequencies stored in Column B). Press (tab) to scroll to the 1st **Result Column** and type **c** (for Column C). Press (tab) to select **OK**. Press (enter). The spreadsheet contains two new columns, Column C and Column D. Column C contains the labels of the numerical values that are displayed in Column D. Use the arrow keys (▲ ▶ ◀ ▼) to move to any location in column D and press (menu). Select **1:Actions, 2:Resize** and **1:Resize Column Width.** Click the right arrow until Column D has been expanded to a larger width. Press (enter). Press the down arrow to remove the shading. Use the down arrow to scroll through the values in Column D.

Describing the Relation between Two Variables

Section 4.1

▸ Example 3 (pg. 198) A Scatter Diagram and Correlation Coefficient

Press STAT, highlight **1:Edit** and clear **L1** and **L2**. Refer to Table 1 on pg. 193. Enter the values of the predictor variable (Club Head Speed) into **L1** and the values of the response variable (Distance) into **L2**. Press **2nd [STAT PLOT]**, select **1:Plot1**, turn **ON** Plot 1 and press ENTER. For **Type** of graph, select the **scatter plot** which is the first selection. Press ENTER. Enter **L1** for **Xlist** and **L2** for **Ylist**. Highlight the first selection, the small square, for the type of **Mark**. Press ENTER. Press ZOOM and 9 to select **ZoomStat**.

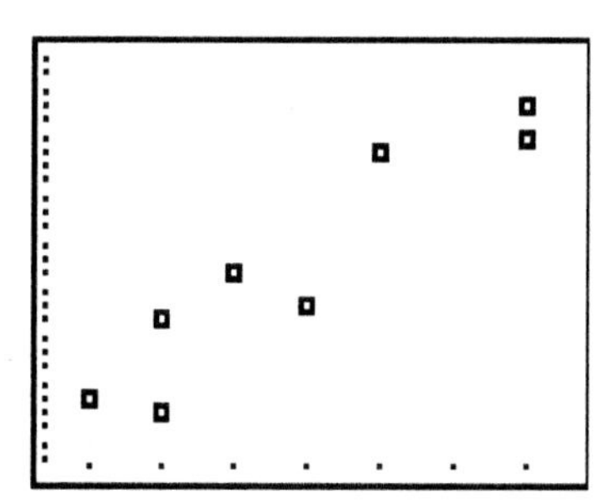

This graph shows a positive linear correlation: as 'Club Head Speed' increases, 'Distance' also increases.

In order to calculate r, the correlation coefficient, you must turn **On** the **Diagnostic** command. Press **2nd [CATALOG]** (Note: **CATALOG** is found above the 0 key). The CATALOG of functions will appear on the screen. Use the down arrow to scroll to the **DiagnosticOn** command.

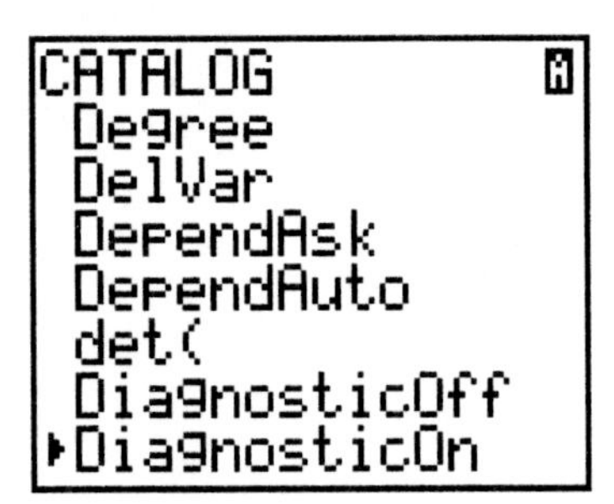

Press ENTER ENTER.

Press STAT, highlight **CALC**, press **4** for **4:LinReg(ax+b)** and press ENTER. (Note: This command gives you the option of specifying which lists contain the X-values and Y-values. If you do not specify these lists, the defaults are used. The defaults are: **L1** for the X-values and **L2** for the Y-values.)

LinReg
y=ax+b
a=3.166101695
b=-55.79661017
r^2=.8811498758
r=.9386958377

The correlation coefficient is r = 0.9386958377. This indicates a strong positive linear correlation between X and Y.

Section 4.2

‣ Example 3 (pg.214) Least Squares Regression Line

Press **STAT**, highlight **1:Edit** and clear **L1** and **L2**. Using Table 1 on pg.193, enter the values of the predictor variable (Club Head Speed) into **L1** and the values of the response variable (Distance) into **L2**. Press **STAT**, highlight **CALC** and press **4** for **4:LinReg(ax+b)**. This command has several options. One option allows you to store the regression equation into one of the Y-variables. To use this option, with the cursor flashing on the line **LinReg(ax+b)**, press **VARS**.

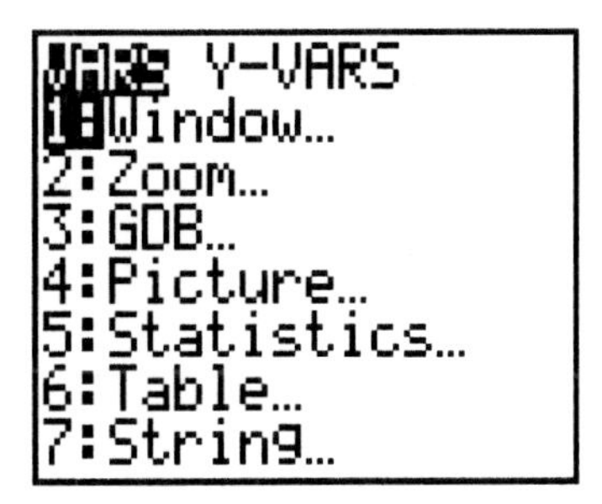

Highlight **Y-VARS.**

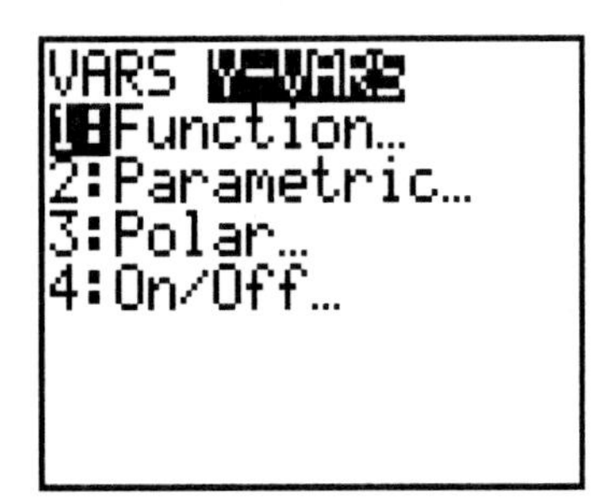

Select **1:Function** and press **ENTER**

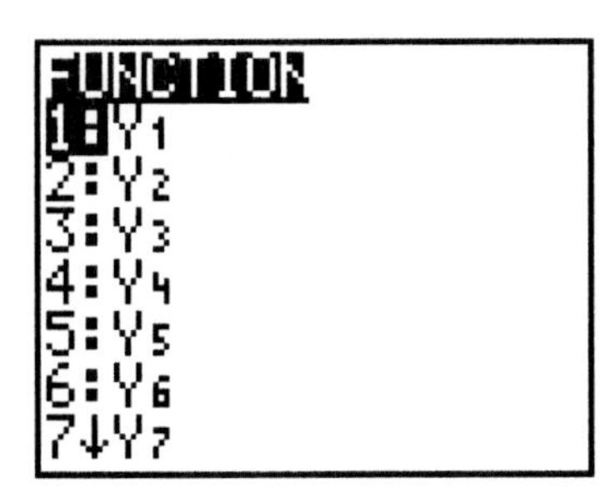

Notice that **1:Y1** is highlighted. Press ENTER.

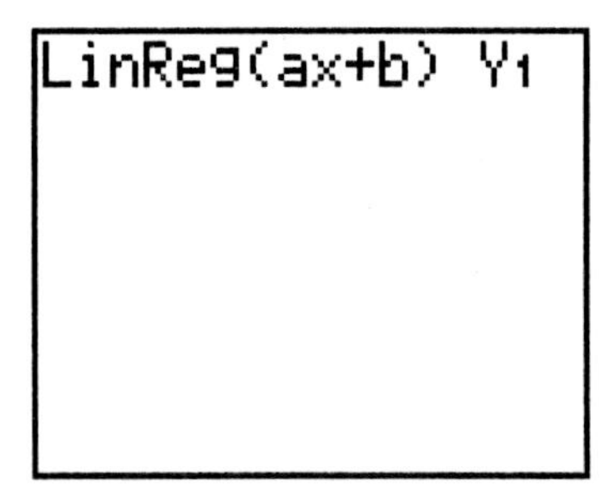

Press ENTER.

```
LinReg
 y=ax+b
 a=3.166101695
 b=-55.79661017
 r²=.8811498758
 r=.9386958377
```

The output displays the general form of the regression equation: y = ax+b followed by values for a and b. Next, r^2, the coefficient of determination, and r, the correlation coefficient , are displayed. If you put the values of a and b into the general equation, you obtain the specific linear equation for this data: $\hat{y}$ = 3.17x + -55.80. Press Y= and see that this specific equation has been pasted to **Y1**.

```
Plot1 Plot2 Plot3
\Y1=3.1661016949
153X+-55.7966101
695
\Y2=
\Y3=
\Y4=
\Y5=
```

Press 2nd **[STAT PLOT]**, select **1:Plot1**, turn **ON** Plot1, select **scatter plot**, set **Xlist** to **L1** and **Ylist** to **L2**. Press ZOOM and 9.

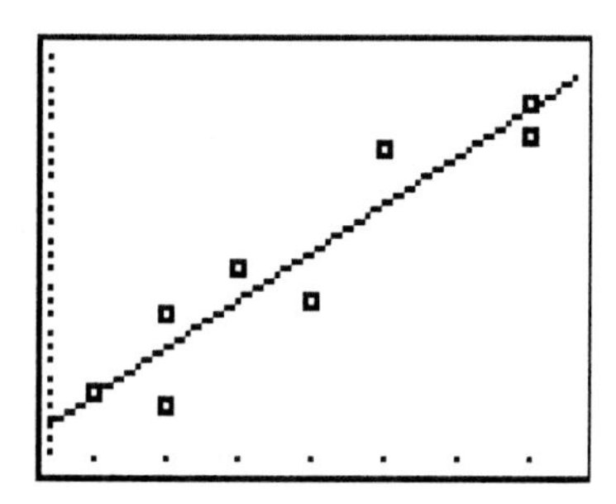

This picture displays a scatter plot of the data and the regression line. The picture indicates a strong positive linear correlation between X and Y, which is confirmed by the r-value of 0.939.

You can use the regression equation stored in **Y1** to predict Y-values for specific X-values. For example, suppose you would like to use the regression equation to predict the 'distance' a golf ball would travel when hit with a 'club head speed' of 103 mph. In other words, for X = 103, what does the regression equation predict for Y? To find this value for $\hat{y}$, press **VARS**, highlight **Y-VARS**, select **1:Function**, press **ENTER**, select **1:Y1** and press **ENTER**. Press **(** 103 **)** and press **ENTER**.

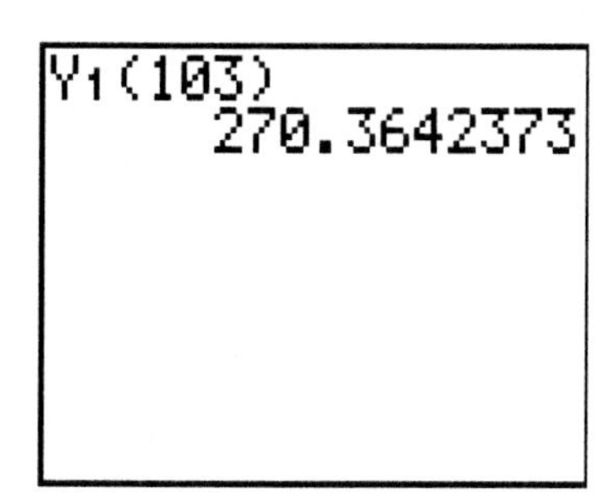

The output shows the predicted Y-value of 270.364 for the input X-value of 103.

The residual is: the actual Y-value – the predicted Y-value: 274 – 270.364 = 3.636.

◀

▸ Problem 22 (pg.219) The Equation of the Regression Line

Enter the predictor variable values into **L1** and the response variable values into **L2**. Press **STAT**. Highlight **CALC**, press **4** for **4:LinReg(ax+b)**,. Press **VARS**, highlight **Y-VARS**, select **1:Function**, press **ENTER** and select **1:Y1** and press **ENTER**.

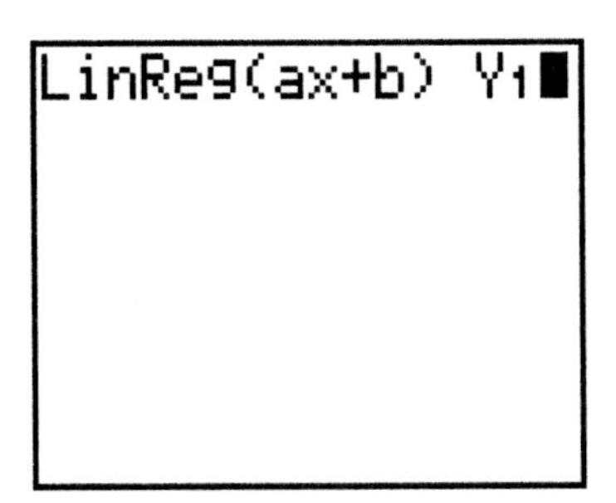

Press **ENTER**.

```
LinReg
 y=ax+b
 a=1.694168031
 b=-142.4709241
 r²=.4954796815
 r=.7039031762
```

(a.) Using **a** and **b** from the output display, the resulting regression equation is $\hat{y}$ = 1.694x – 142.471. Press **Y=** to confirm that the regression equation has been stored in **Y1**. Press **2nd [STAT PLOT]**, select **1:Plot1**, turn **ON** Plot1, select **scatter plot**, set **Xlist** to **L1** and **Ylist** to **L2**. Press **ZOOM** and **9** and a graph of the scatter plot with the regression line will be displayed.

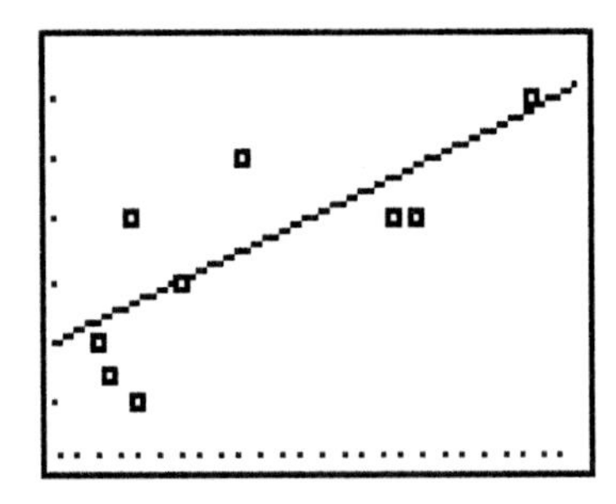

(c.) Next, you can use the regression equation to predict the weight of a 149.0 cm bear. Press VARS, highlight **Y-VARS**, select **1:Function** and press ENTER. Select **1:Y1** and press ENTER. Press (149.0) and ENTER.

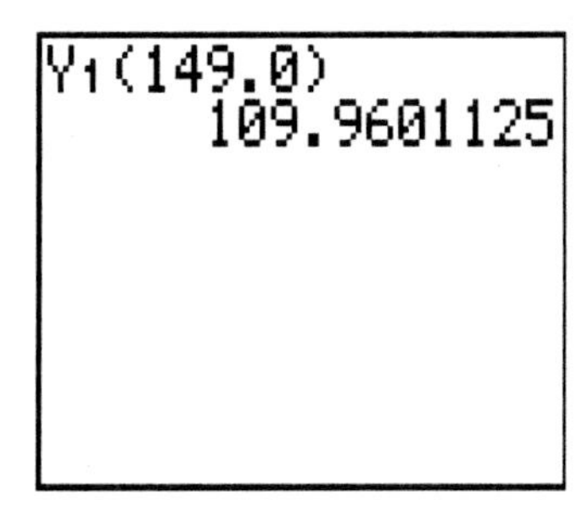

The predicted weight for this bear is 110.0 kg.

(d) Residual = Actual weight – predicted weight: 85 – 110 = -25 kg.

Section 4.3

▸ Example 2 (pg.226) Coefficient of Determination, R^2

Using Table 1 on pg. 193, enter the X-values into **L1** and the Y-values into **L2.** Press **STAT**. Highlight **CALC**, select **4:LinReg(ax+b)**, press **ENTER**. (For this example, we are not storing the regression equation in **Y1.**) .

```
LinReg
 y=ax+b
 a=3.166101695
 b=-55.79661017
 r²=.8811498758
 r=.9386958377
```

The value of r^2, .881, is displayed in the output.

◂

Example 3 (pg. 228) Is a Linear Model Appropriate?

Using the data in Table 7, enter the X-values into **L1** and the Y-values into **L2**. Press STAT, highlight **CALC** and press **4** for **4:LinReg(ax+b)** and press ENTER

```
LinReg
 y=ax+b
 a=-.7928571429
 b=165.65
 r²=.9933086101
 r=-.9966486894
```

To plot the residuals, first make sure that there is nothing stored in the Y-registers. Press Y= and check the Y-registers. If any of them contain a function, move the cursor to that Y-register and press CLEAR.

Press **2nd** **[STAT PLOT]**, select **1:Plot1**, turn **ON** Plot 1 and press ENTER. For **Type** of graph, select the **scatter plot** which is the first selection. Press ENTER. Enter **L1** for **Xlist.** Move the cursor to **Ylist**. Press 2nd **[List]** and select **Resid**. Highlight the first selection, the small square, for the type of **Mark**. Press ENTER. Press ZOOM and 9 to select **ZoomStat**.

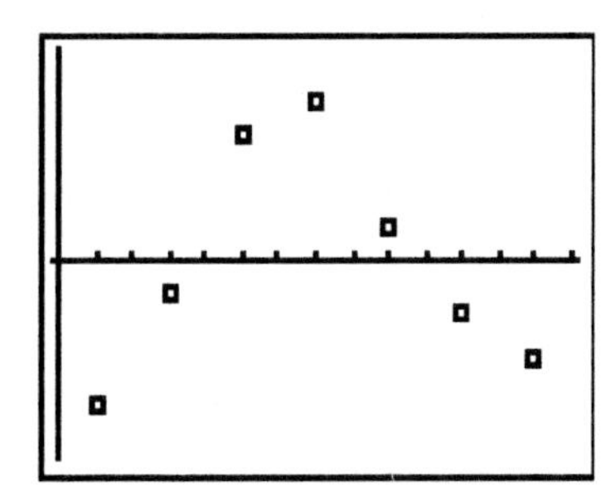

This graph of the residuals vs. the predictor variable (Time) shows a pattern (an upside-down U-shape), which indicates that the linear model is not appropriate.

‣ Example 6 (pg. 230) Graphical Residual Analysis

Using the data in Table 5 on pg.217, enter the X-values, (Club Head Speed), into **L1** and the Y-values, (Distance), into **L2**. Press STAT, highlight **CALC** and select **4:LinReg(ax+b)** and press ENTER

To plot the residuals, first make sure that there is nothing stored in the Y-registers. Press Y= and check the Y-registers. If any of them contain a function, move the cursor to that Y-register and press CLEAR.

Press **2nd [STAT PLOT]** , select **1:Plot1**, turn **ON** Plot 1 and press ENTER. For **Type** of graph, select the **scatter plot** which is the first selection. Press ENTER. Enter **L1** for **Xlist.** Move the cursor to **Ylist**. Press 2nd **[List]** and select **Resid**. Highlight the first selection, the small square, for the type of **Mark**. Press ENTER. Press ZOOM and 9 to select **ZoomStat**.

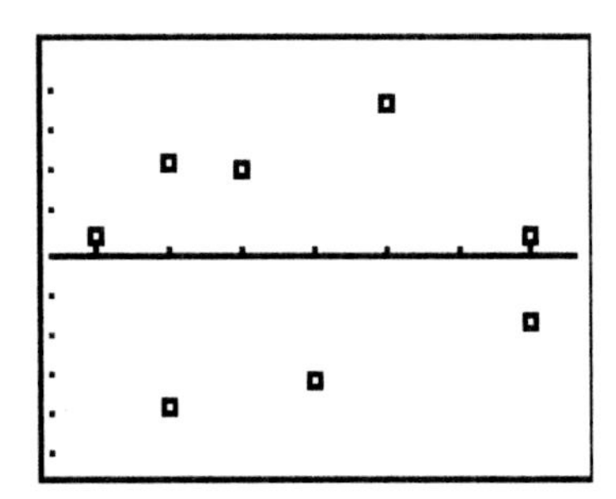

There is no discernable pattern in the plot of the residuals. This indicates support for the linear model that we used to calculate the regression equation.

The next step in analyzing the residuals is to construct a boxplot to determine if there are any unusual residuals values (called outliers.)

Press **2nd [STAT PLOT]**, select **1:Plot1**, turn **ON** Plot 1 and press ENTER. For **Type** of graph, select the **boxplot** with outliers, which is the first selection in the second row. Press ENTER. Move the cursor to **Xlist**. Press 2nd **[List]** and select **7:Resid**. Move the cursor to **Freq** and set this equal to **1**. Highlight the first selection, the small square, for the type of **Mark**. Press ZOOM and 9 to select **ZoomStat**.

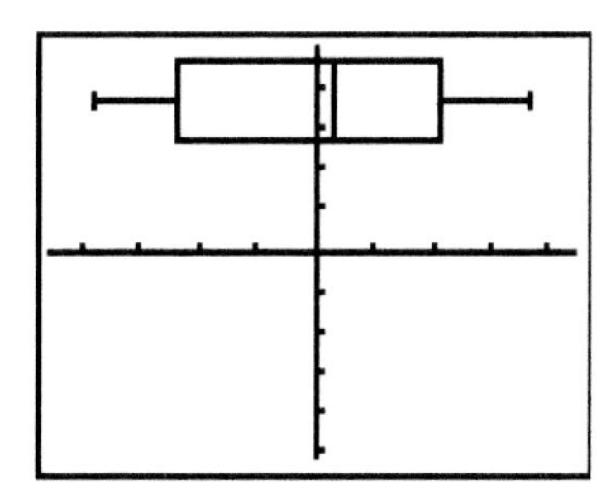

If you would like to remove the X and Y axes from the boxplot press **2nd [FORMAT]** (found above the **ZOOM** key). Scroll down to fourth line, use the right arrow to highlight **AxesOff** and press ENTER. Press ZOOM and 9 to select **ZoomStat**.

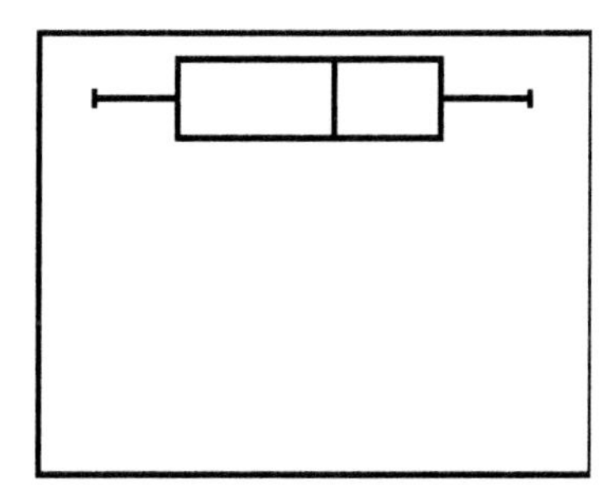

In this example, there are no outliers. This further supports the use of the linear model that was selected for this dataset.

▸ Problem 29 (pg. 235)

Enter the X-values (distance from the sun) into **L1** and the Y-values (sidereal year) into **L2**.

a.) First make sure that there is nothing stored in the Y-registers. Press **Y=** and check the Y-registers. If any of the Y-registers contain a function, move the cursor to that Y-register and press **CLEAR**.

Press **2nd [STAT PLOT]** , select **1:Plot1**, turn **ON** Plot 1 and press **ENTER**. For **Type** of graph, select the **scatter plot** which is the first selection. Press **ENTER**. Enter **L1** for **Xlist** and **L2** for **Ylist**. Highlight the first selection, the small square, for the type of **Mark**. Press **ENTER**. Press **ZOOM** and ▼ to select **ZoomStat**.

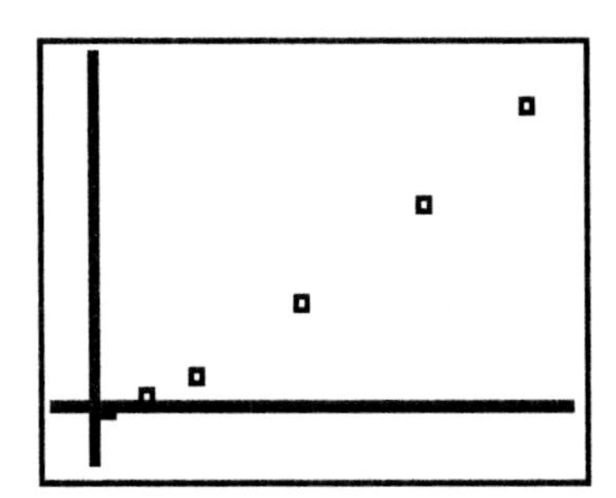

(Note: It is difficult to see all the nine points on this graph. It looks as if there are only 6 data points. In fact, there are nine points but, the first three points are so close together that they are indistinguishable from one another.)

b. and c) Press **STAT**, highlight **CALC** and press **4** for **4:LinReg(ax+b)** and press **ENTER**

```
LinReg
 y=ax+b
 a=.0656909572
 b=-12.49670982
 r²=.9779211075
 r=.9888989369
```

d.) Press **2nd [STAT PLOT]** , select **1:Plot1**, turn **ON** Plot 1 and press **ENTER**. For **Type** of graph, select the **scatter plot** which is the first selection. Press **ENTER**. Enter **L1** for **Xlist.** Move the cursor to **Ylist**. Press 2nd **[List]** and select **7:Resid**. Highlight the first selection, the small square, for the type of **Mark**. Press ENTER. Press ZOOM and 9 to select **ZoomStat**.

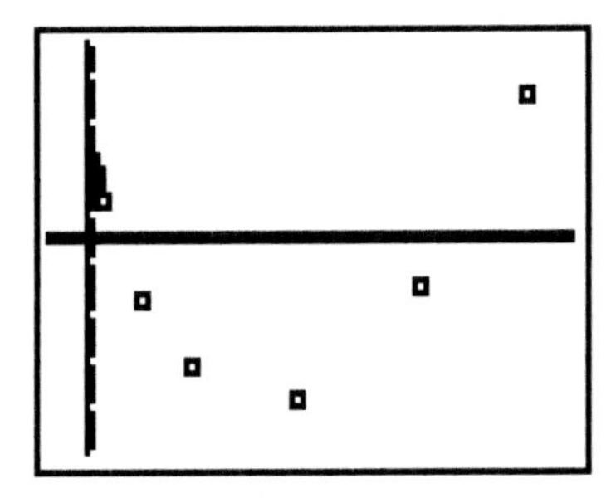

e.) This graph of the residuals vs. the x- variable shows a U-shaped pattern, which indicates that the linear model is not appropriate.

◀

‣ Problem 34 (pg.236)

Enter the X-values (heights) into **L1** and the Y-values (weights) into **L2**.

To analyze the data value for Derek Jeter, we will calculate and plot the regression equation and then plot the residuals.

Press **STAT**, highlight **CALC** and press **4** for **4:LinReg(ax+b)** and press **VARS**, highlight **Y-VARS**, select **1:Function**, press **ENTER** and select **1:Y1** and press **ENTER** **ENTER**

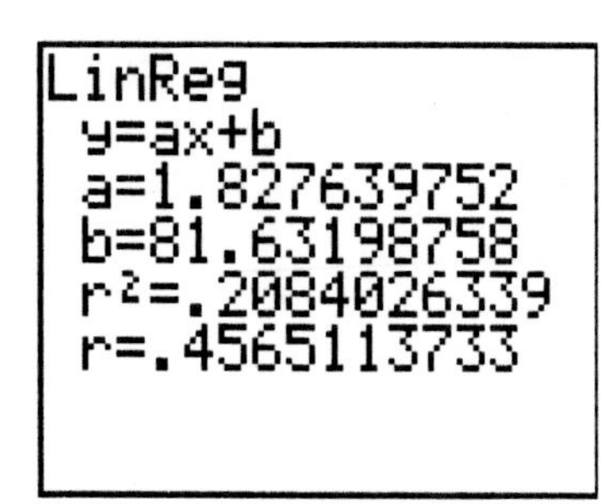

Press **2nd** **[STAT PLOT]**, select **1:Plot1**, turn **ON** Plot1, select **scatter plot**, set **Xlist** to **L1** and **Ylist** to **L2**. Press **ZOOM** and **9**.

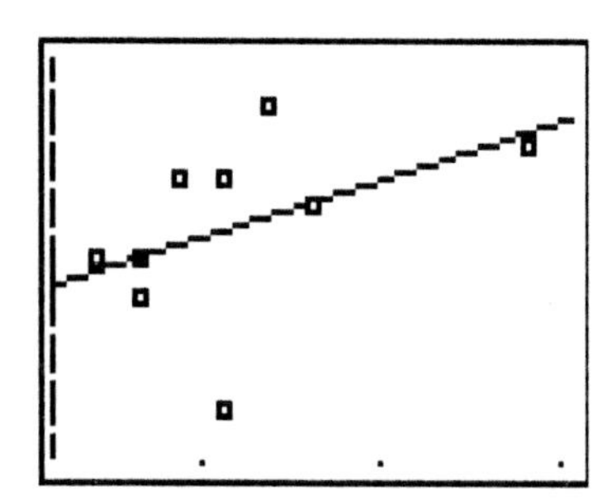

Press **TRACE** and use the right arrow to scroll through the data. Notice that Derek Jeter is the data point that is the farthest from the regression line.

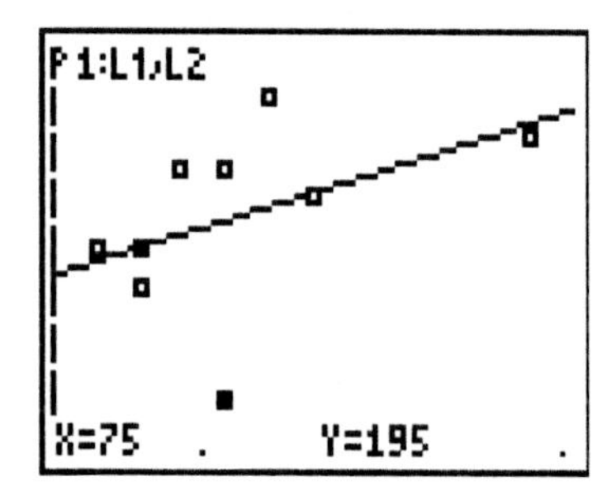

Next, look at the residual plot. To plot the residuals, first make sure that there is nothing stored in the Y-registers. Press **Y=** and check the Y-registers. If any of them contain a function, move the cursor to that Y-register and press **CLEAR**.

Press **2nd [STAT PLOT]** , select **1:Plot1**, turn **ON** Plot 1 and press **ENTER**. For **Type** of graph, select the **scatter plot** which is the first selection. Press **ENTER**. Enter **L1** for **Xlist.** Move the cursor to **Ylist**. Press 2nd **[List]** and select **7:Resid**. Highlight the first selection, the small square, for the type of **Mark**. Press **ENTER**. Press **ZOOM** and **9** to select **ZoomStat**.

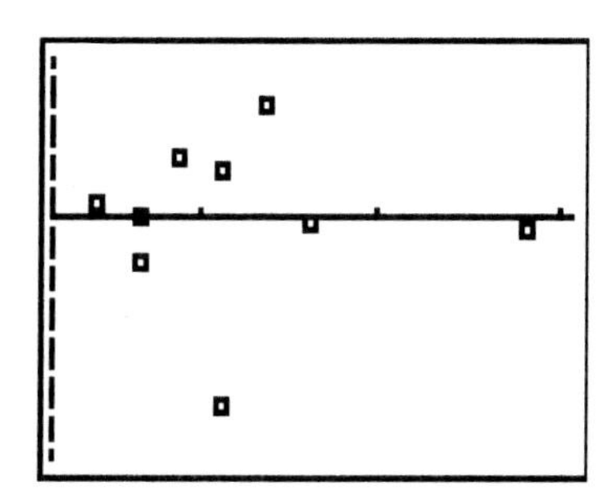

Press **TRACE** and use the right arrow to scroll through the data. Notice that Derek Jeter is the data point that is the farthest from the horizontal line.

The next step in analyzing the residuals is to construct a boxplot to determine if there are any unusual residuals values (called outliers.)

Press **2nd [STAT PLOT]**, select **1:Plot1**, turn **ON** Plot 1 and press **ENTER**. For **Type** of graph, select the **boxplot** with outliers, which is the first selection in the second row. Press **ENTER**. Move the cursor to **Xlist**. Press 2nd **[List]** and select **7:Resid**. Move the cursor to **Freq** and set this equal to **1**. Highlight the first selection, the small square, for the type of **Mark**. Press **ZOOM** and **9** to select **ZoomStat**.

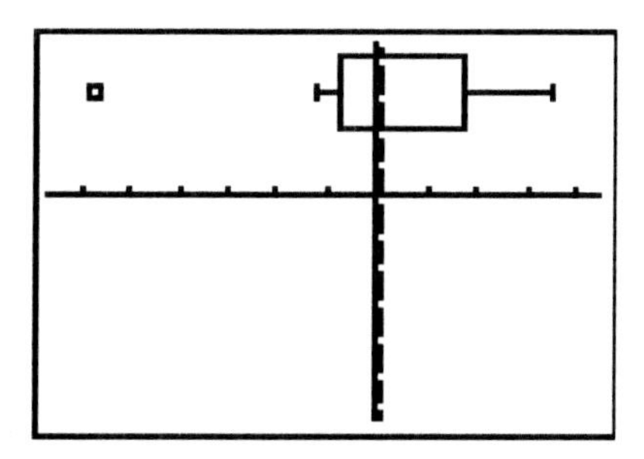

Notice that there is one outlier. This is the data point for Derek Jeter. The analysis confirms that Derek Jeter is significantly different from the rest of the data. His weight is unusually low for his height.

◀

Section 4.5 (Note: This Section is available on CD)

▸ Example 4, pg. 4-3 An Exponential Model

The TI-84 has the capability of creating an exponential model directly from the data. It is not necessary to transform the data by taking the log of the y-values.

a.) Enter the times into **L1** and the temperatures into **L2**. Make sure that there is nothing stored in the Y-registers. Press Y= and check the Y-registers. If any of them contain a function, move the cursor to that Y-register and press CLEAR. Press **2nd [STAT PLOT]**, select **1:Plot1**, turn **ON** Plot 1 and press ENTER. For **Type** of graph, select the **scatter plot** which is the first selection. Press ENTER. Enter **L1** for **Xlist** and **L2** for **Ylist**. Highlight the first selection, the small square, for the type of **Mark**. Press ENTER. Press ZOOM and 9 to select **ZoomStat.**

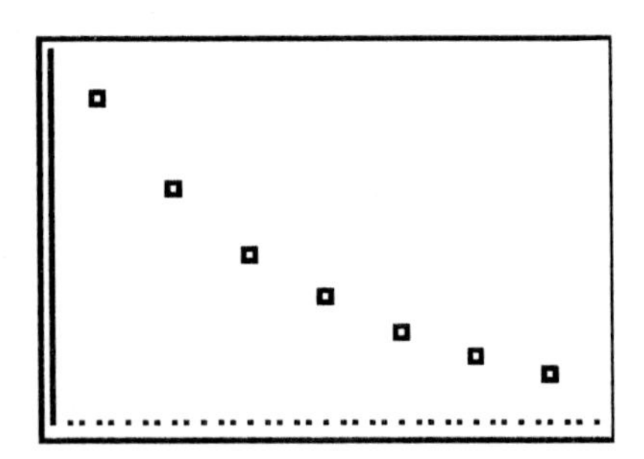

The shape of the scatter plot suggests that an exponential model would be appropriate for this dataset.

(d.) Press STAT, highlight **CALC** and select **0:ExpReg.** Press VARS, highlight **Y-VARS**, select **1:Function**, press ENTER and select **1:Y1** and press ENTER ENTER.

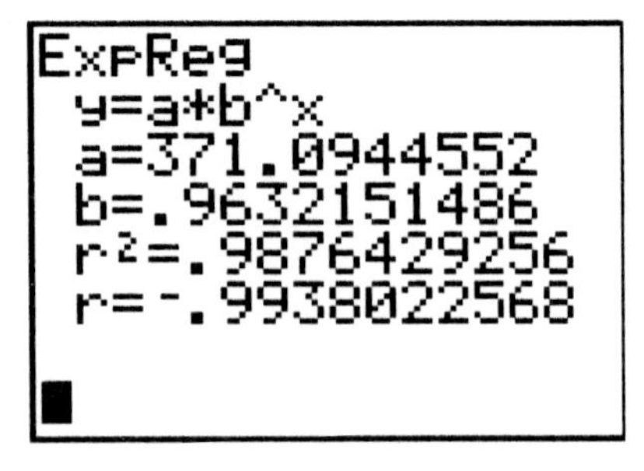

The exponential equation for this dataset is: $\hat{y} = 371.0906 * (0.9632)^x$.
If you press GRAPH, you can see a picture of the data along with the exponential model of best fit.

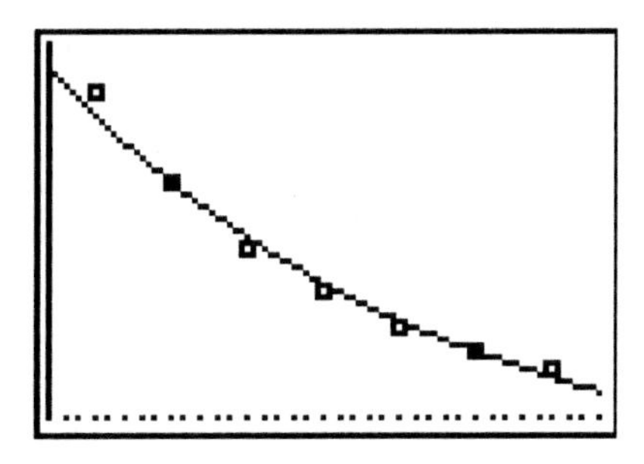

e.) To predict the temperature at a time of 40 minutes, press VARS, highlight **Y-VARS**, select **1:Function** and press ENTER. Select **1:Y1** and press ENTER. Press (40) and ENTER.

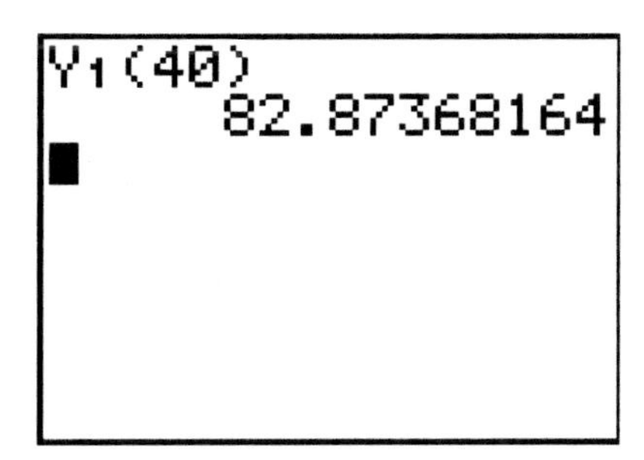

▸ Example 5, pg. 4-6 A Power Model

The TI-84 has the capability of creating a power model directly from the data. It is not necessary to transform the equation by taking the log of the y-values. (Note: This model can used only in situations where x > 0, y > 0.)

a.) Enter the X-values into **L1** and the Y-values into **L2**. Make sure that there is nothing stored in the Y-registers. Press Y= and check the Y-registers. If any of them contain a function, move the cursor to that Y-register and press CLEAR. Press **2nd [STAT PLOT]** , select **1:Plot1**, turn **ON** Plot 1 and press ENTER. For **Type** of graph, select the **scatter plot** which is the first selection. Press ENTER. Enter **L1** for **Xlist** and **L2** for **Ylist**. Highlight the first selection, the small square, for the type of **Mark**. Press ENTER. Press ZOOM and 9 to select **ZoomStat.**

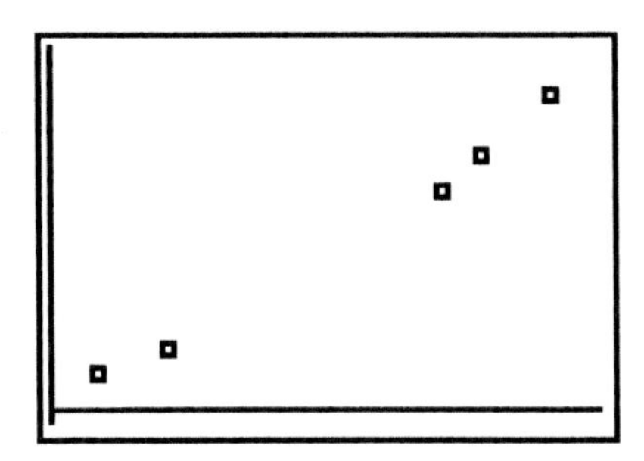

This is a very small dataset and it is difficult to determine which model would be the most appropriate one. For the purposes of this example, we will select the **power model**.

(d.) Press STAT, highlight **CALC** and scroll down to **A:PwrReg** and press ENTER. Press VARS, highlight **Y-VARS**, select **1:Function**, press ENTER and select **1:Y1** and press ENTER.

```
PwrReg
 y=a*x^b
 a=4.933897526
 b=1.992836932
 r²=.9999944648
 r=.9999972324
```

The power equation for this dataset is: $y = 4.934(x)^{1.99284}$.

If you press GRAPH , you can see a picture of the data along with the power model of best fit.

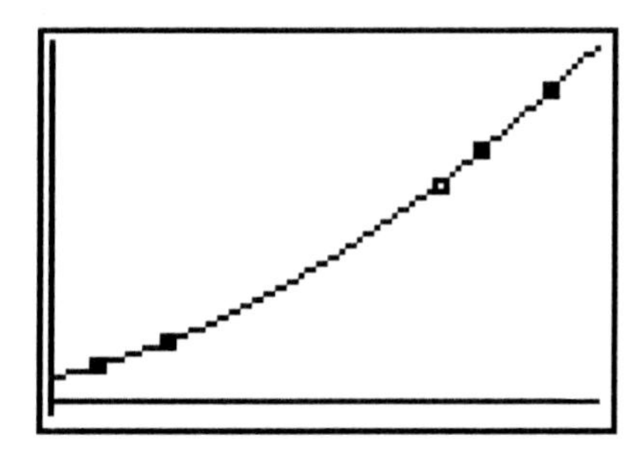

e.) To predict the distance a ball would have fallen if it took 4.2 seconds to hit the ground, press VARS, highlight **Y-VARS**, select **1:Function** and press ENTER. Select **1:Y1** and press ENTER. Press (4.2) and ENTER.

```
Y1(4.2)
       86.14386016
```

◀

TI-89 Instructions:

These instructions are designed to give you an overview of the regression capabilities of the TI-89.

Select the **Stat/List** icon. The screen that follows allows you to store the data in specific folders. It is easy to always use the **Main** folder so simply press **ENTER**. The next screen displays Lists 1 through 4. If any Lists contain data and you want to delete that data, press **F3**, select **1:Names** and scroll down to each List that you would like to delete. While the list is highlighted, press **F4** to select it. Once you have selected all the lists that you want to delete, press **F1** and select **1:Delete** and press **ENTER**. Press **ESC** to return to the lists. Notice that the lists that you just deleted are no longer displayed. Press **F1** and select **3:Setup Editor** and press **ENTER**. Lists 1 through 4 are again displayed. Now you are ready to enter the data. Enter the x-values into **List1** and enter the y-values into **List2**.

Press **F4** and select **3:Regressions**. There are several different equations that you can select. The first two choices are linear regression equations. Your textbook uses the linear regression equation, y = ax+b, which is selection **2**. On the next screen, use **2nd** **VAR-LINK** and choose **list1** for **XList** and **list2** for **YList**. On the next line, '**Store RegE to**,' select **y1(x)**. Press **ENTER** and the regression output will be displayed. Press **ENTER** again and a column of residuals (**resid**) will be displayed in the **List Editor**.

To display the scatterplot with the regression equation, press **F2** and select **1:Plot Setup**. Press **ENTER**. Press **F1** to **Define your plot**. On the Plot Setup screen, use the right arrow to display the **Plot Type** options. Select **scatter** and fill in the required entry boxes. Click **ENTER** until you return to the **Plot Setup** screen and press **F5** to display the graph.

To display a plot of the residuals, first deselect y1(x) by pressing ◆ and **Y=** and then pressing **F4.** Press **2nd** **QUIT**. Return to the data lists in the Stats/List editor and go through the steps to set up Plot 1. You need to change the list selected for the Y-variable to the list containing the residuals. To do this, press **2nd** **VAR-LINK**. On the **Var-Link** screen, press **F2**. On the next screen, on the 2nd line, **Folder**, use the right arrow and select **5:Statvars**. On the next line, **Var type**, use the right arrow and select **3:List**. Press **ENTER**. The **resid** list will be displayed. Highlight it and press **ENTER** to select it. Click **ENTER** until you return to the Plot Setup page and press **F5.** A plot of the residuals will be displayed.

TI-*n*spire Instructions:

These instructions are designed to give you an overview of the regression capabilities of the TI-*n*spire handhelds.

Press (home) and select **6:New Document**. (Note: If you currently have a document open, the next screen will ask if you want to save the document. Press (tab) to select **No**. Press (enter).) Select **3:Add Lists & Spreadsheet.** Move to the top of column **A** and in the box next to '**A**' type in a name. Move to **Line 1** in column **A** and begin entering your **x-values**. Move to the top of column **B** and in the box next to '**B**' type in a name. Move to **Line 1** in column **B** and begin entering your **y-values**. Press (menu), select **4:Statistics** and select **1: Stat Calculations.** There are several different regression equations from which to choose. For a linear regression of the form **(a+bx)** select **4**. On the next screen, press the down arrow, ▼, and highlight the name you are using for column **A**. Press (enter). Press (tab) to move to **YList**. Use the down arrow and highlight the name you are using for Column **B**. Press (enter). Press (tab). You can use the default name to '**Save RegEqn**.' Tab to the entry: **1st Result Column** and type in '**c**' for Column **C**. Press (enter). Press (tab) to highlight **OK** and press (enter). The regression output is displayed in Columns **C** and **D**.

For easier viewing of the output, you can expand columns **C** and **D**. To expand a column, move to any position in the column. Press (menu), select **1:Actions**, **2:Resize** and **1:Resize Column Width**. Use the right arrow, ▶, to expand the column width. Press (enter). Press the down arrow, ▼, to remove the highlighting.

To display the scatterplot with the regression line, press (home) and select **5:Data & Statistics**. A new page with a scatter of points will be displayed. Press and hold the down arrow to move the cursor to the middle of the x-axis at the bottom of the screen. When the box '**Click to add variable**' appears, press (enter). Highlight the Column **A** name and press (enter). Use the arrow keys to move the cursor to the middle of the y-axis on the left side of the screen. When the box '**Click to add variable**' appears, press (enter). Highlight the Column **B** name and press (enter). A scatterplot of the data will appear. To add the regression line, press (menu), select **3:Actions,** select **5: Regression** and **2:Show Linear (a+bx).**

To display a plot of the residuals, begin by removing the regression line: press (menu), select **3:Actions,** select **5: Regression** and **2:Hide Linear (a+bx).** Use the arrow keys to move to the y-axis. When the box '**Click to change variable**' appears, press (enter). Highlight the name '**stat.resid**' and press (enter).

CHAPTER 5

Probability

Section 5.1

Example 8 (pg. 267) Simulating Probabilities

In this example, we will use simulation to estimate the probability that a three-child family has two boys. We assume the simple events, "having a boy," and "having a girl," are equally likely. In this simulation, we will designate "0" as a "girl," and "1" as a "boy."

a.) The first step is to set the *seed* by selecting a 'starting number' and storing this number in **rand**. Suppose, for this example, that we select the number '1970' as the starting number. Type **'1970'** into your calculator and press the **STO** key. Next press the **MATH** key, move the cursor to highlight **PRB**. Select **rand,** and press **ENTER**. The starting value of '1970' will be stored into **rand** and will be used as the *seed* for generating random numbers. (Note: This process of setting the *seed* is optional. You can omit it and simply go directly to the next step.)

(a.) To simulate 100 three-child families, you will start by simulating the gender of the 1st child in each family and store the results in L1. Press **MATH**, highlight **PRB**, and select **5:randInt(** and press **ENTER**. The **randInt(** command requires a minimum value, (which is 0 for this simulation), a maximum value (which is 1), and the number of trials (100). In the **randInt(** command type in **0** , **1** , **100.**

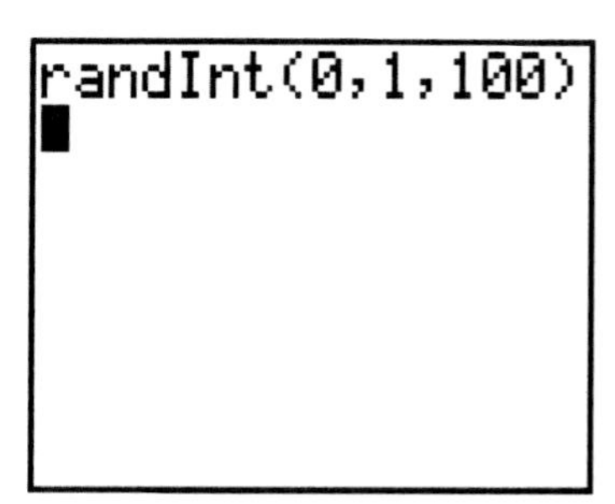

Press ENTER. It will take a few seconds for the calculator to generate 100 numbers. Notice, in the upper right hand corner a flashing | |, indicating that the calculator is working. When the simulation has been completed, a string of **0's** and **1's** will appear on the screen followed by **….**, indicating that there are more numbers in the string that are not shown.

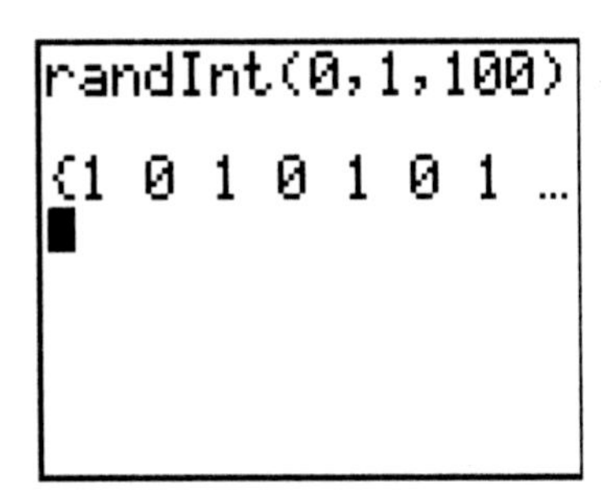

Press STO and **2nd** [**L1**] ENTER. This will store the string of numbers in **L1**.

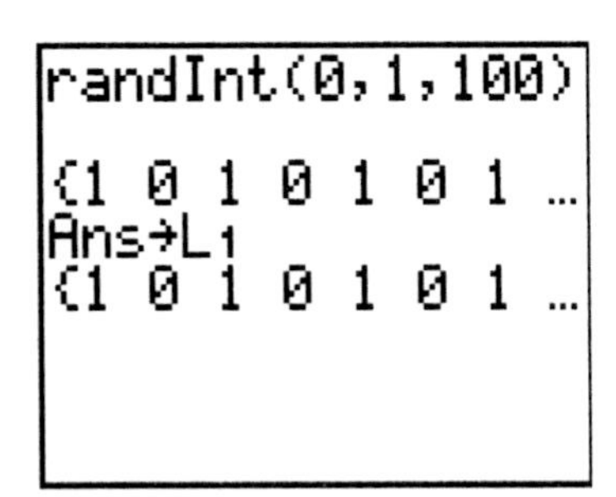

Repeat the process and simulate the gender of the 2nd child in each family and store the results in **L2**. Repeat the process one more time to simulate the gender of the 3rd child in each family and store the results in **L3**.

Next, you will count the number of boys in each family and store the results in **L4**. Since you designated '1' to represent a boy in the simulation, you can add the values in **L1, L2,** and **L3** to count the number of boys in each family. Press STAT and select **1:Edit**. Move the cursor to the top of **L4** (so that the label, **L4**, is highlighted) and press ENTER. On the data entry line, type in **2nd L1+2nd L2+2nd L3** and press ENTER. The numbers in **L4** represent the number of boys in each family.

To find the simulated probability of having two boys in a three-child family, you need to calculate the proportion of times that "two boys" occurred among the 100 families. One way to count the number of 2's in your simulation is to create a histogram of the results. First make sure that there is nothing stored in the Y-registers. Press Y= and check the Y-registers. If any register contains a function, move the cursor to that Y-register and press CLEAR.

To set up the histogram, push **2nd** [STAT PLOT] and ENTER to select **Plot 1**. Turn ON **Plot 1**, set **Type** to **Histogram**, set **Xlist** to **L4,** set **Freq** to **1.** Press Window to adjust the Graph Window. Set **Xmin** equal to 0 (the minimum value in your simulation) and **Xmax** equal to 4 (a value that would be one integer larger than the maximum value (which would be 3 boys). This extra value is needed to complete the last bar of the histogram). Set **Xscl** equal to 1. Set **Ymin** to a small negative value (a value between –6 and –1 would be good). **Ymax** must be larger than the largest frequency value in your dataset. A good value for **Ymax** would be 50. You never need to adjust **Yscl**.

Press GRAPH and the histogram should appear.

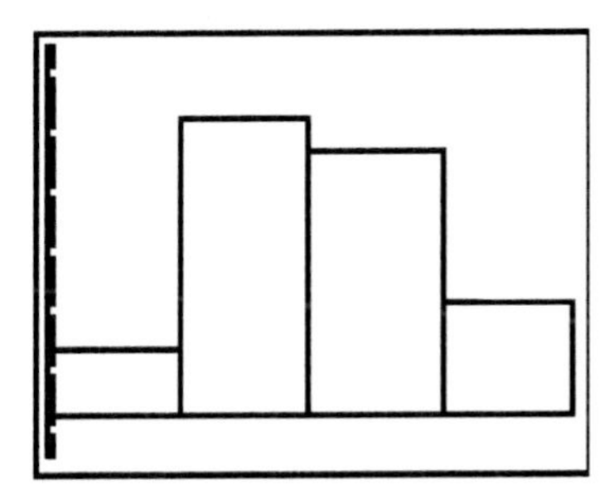

You can press TRACE and scroll through the bars of the histogram. The third bar of the histogram represents all the families with 2 boys.

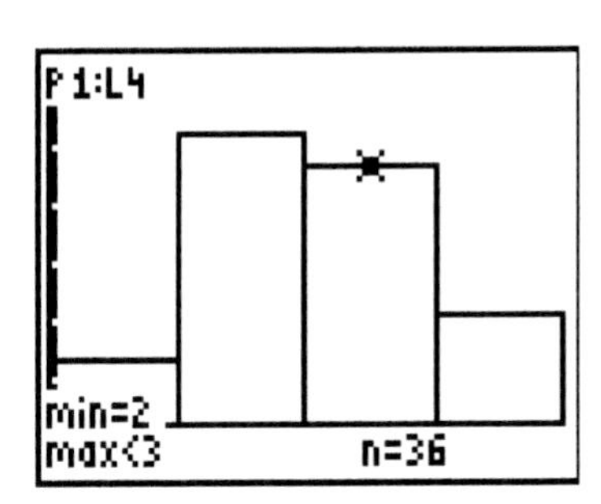

So, based on this simulation, we *estimate* the probability that a three-child family has 2 boys is 36 out of 100 or 36%.

(b.) Repeat the steps in part (a.) and increase the sample size to 999. This is the maximum sample size that the calculator will allow.

◀

Problem 51 (pg 272)

In this simulation, we will use the integers 1,2,3,4,5 and 6 to represent the six possible outcomes on a six-sided die. Press MATH, highlight **PRB**, and select **5:randInt(** and press ENTER. The **randInt(** command requires a minimum value, (which is 1 for this simulation), a maximum value (which is 6), and the number of trials (100). In the **randInt(** command type in **1** , **6** , **100.**

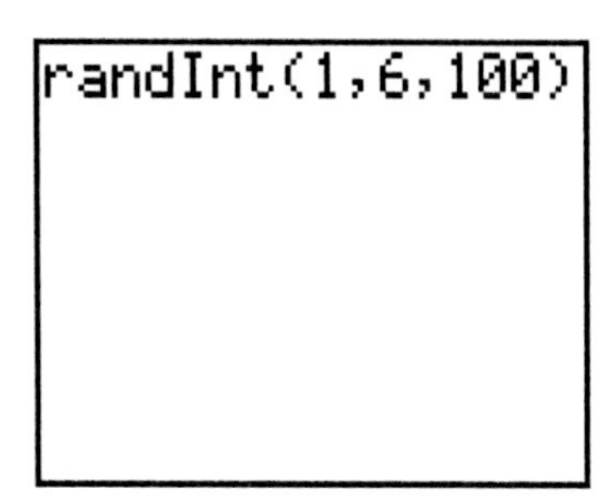

Press ENTER. It will take several seconds for the calculator to generate 100 rolls of the die. Notice, in the upper right hand corner a flashing | , indicating that the calculator is working. When the simulation has been completed, a string of **0's, 1's, 2's. etc.** will appear on the screen followed by **....**, indicating that there are more numbers in the string that are not shown.

Store the data in **L1** by pressing STO and 2nd **[L1]** ENTER.

(a.) One way to count the number of 1's in your simulation is to create a histogram of the results. First make sure that there is nothing stored in the Y-registers. Press Y= and check the Y-registers. If any register contains a function, move the cursor to that Y-register and press CLEAR.

To set up the histogram, push **2nd** [STAT PLOT] and ENTER to select **Plot 1**. Turn ON **Plot 1**, set **Type** to **Histogram**, set **Xlist** to **L1**,. set **Freq** to **1.** Press Window to adjust the Graph Window. Set **Xmin** equal to 1 (the minimum value in your simulation) and **Xmax** equal to 7 (a value that would be one integer larger than the maximum value on the roll of a die. This extra value is needed to complete the last bar of the histogram). Set **Xscl** equal to 1. Set **Ymin** to a small negative value (a value between –6 and –1 would be good). **Ymax** must be larger than the largest frequency value in your dataset. A good value for **Ymax** would be 30. You never need to adjust **Yscl**.

Press GRAPH and the histogram should appear.

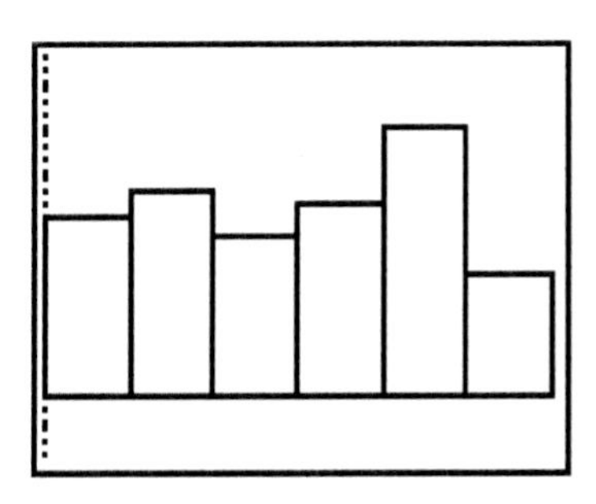

You can press **TRACE** and scroll through the bars of the histogram.
The first bar of the histogram represents all the rolls that resulted in 1's.

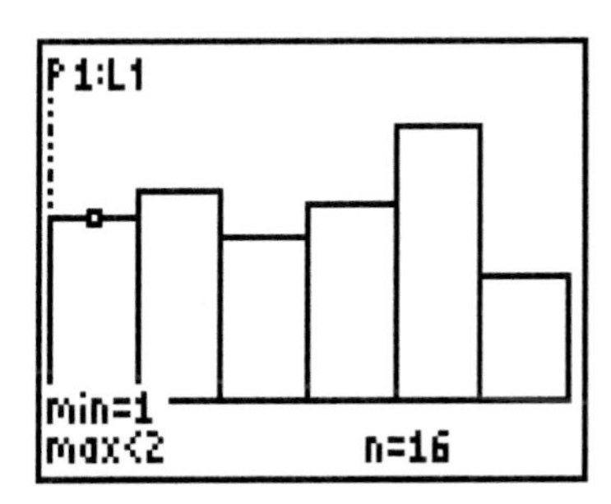

Notice, for this example, there were 16 rolls of 1's. So, based on this simulation, the estimated probability of rolling a '1' is 16 out of 100 or 16%.

(b.) Repeat the simulation and create a new histogram.

(c.) Repeat the simulation and increase the number of rolls of the die to 500 and create a new histogram. (Note: Increase the value of **Ymax** to 100 for this histogram.)

◀

Section 5.5

Example 4 (pg. 305) The Traveling Salesman - Factorials

The total number of different routes that are possible can be computed using the factorial function.

Press 7, MATH, highlight **PRB** and select **4:!** and ENTER.

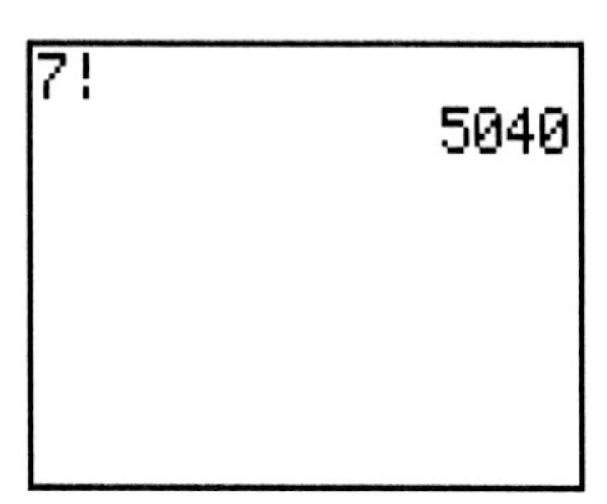

There are 5,040 different possible routes.

Example 6 (pg. 306) Permutations

(a.) In this example, there are 7 objects (n=7). From these 7 objects, 5 objects are selected (r=5). The permutation formula counts the number of different ways that these 5 objects can be selected and arranged from the total of 7 objects. The formula **nPr** is used with **n = 7** and **r =5.** So, the formula is **7P5**.

Enter the first value, **7**, and press **MATH**, highlight **PRB** and select **2:nPr** and **ENTER**.

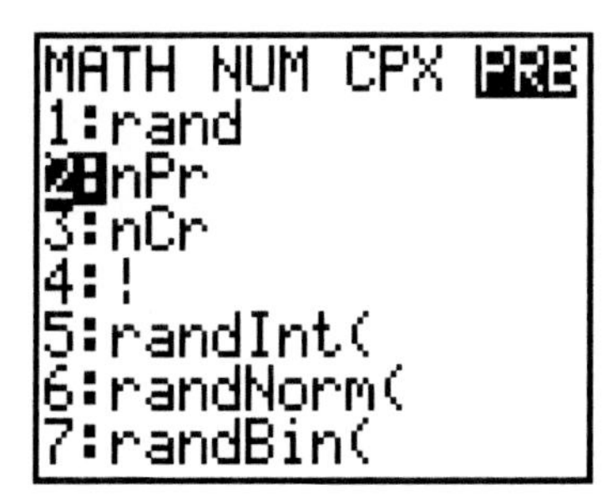

Now enter the second value, **5**, and **ENTER**. The answer, 2520, appears on the screen.

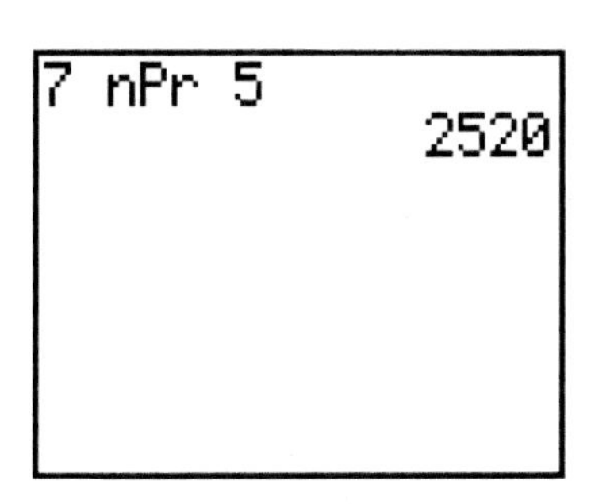

▸ Example10 (pg.309) Combinations

(b.) In this example, there are 6 objects (n=6). From these 6 objects, 4 objects are selected (r=4). The combination formula counts the number of different ways that these 4 objects can be selected from the total of 6 objects. The formula **nCr** is used with **n =6** and **r =4.** So, the formula is **6C4**.

Enter the first value, **6**, and press **MATH**, highlight **PRB** and select **3:nCr** and **ENTER**.

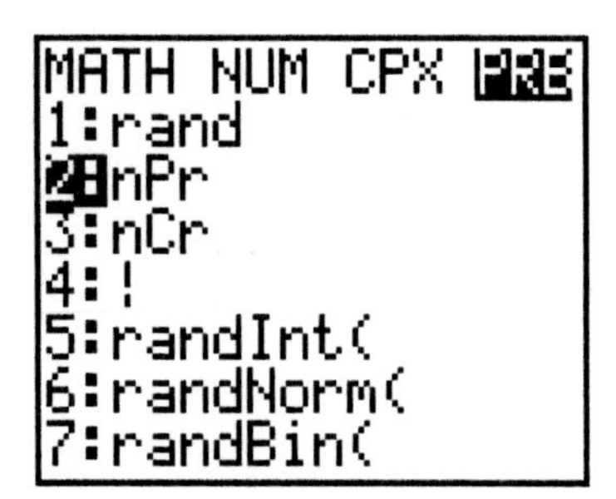

Enter the second value, **4**, and **ENTER**. The answer, 15, appears on the screen.

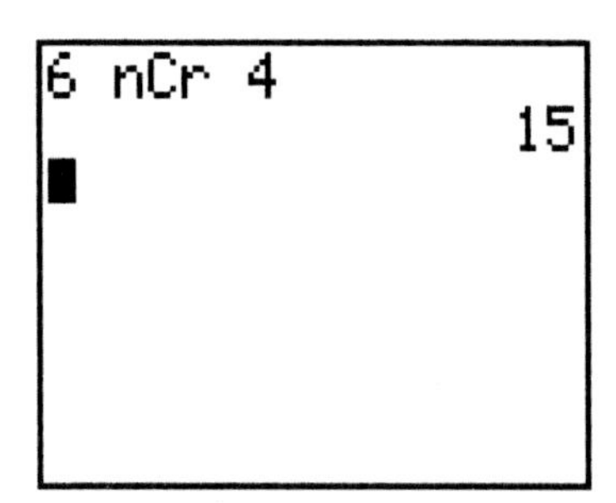

◀

Example 13 (pg. 311) Arranging Flags - Permutations with Non-distinct Items

To calculate $\frac{10!}{5!3!2!}$ you will use the factorial function (**!**). Enter the first value, **10**, press **MATH**, highlight **PRB** and select **4:!** . Then press **÷**. Open the parentheses by pressing **(** . Enter the next value, **5**, press **MATH**, **PRB**, and select **4:!** . To multiply by 3**!**, press **x** and enter the next value, **3**. Press **MATH**, **PRB**, and select **4:!** . To multiply by 2**!**, press **x** and enter the next value, **2**. Press **MATH**, **PRB**, and select **4:!** . Close the parentheses **)** and press **ENTER**.

```
10!/(5!*3!*2!)
            2520
```

Example 14 (pg. 311) Winning the Lottery - Probabilities Involving Combinations

To calculate the probability of winning the Illinois Lottery, you must calculate $\frac{2}{{}_{52}C_6}$.

Enter the numerator, **2**, into your calculator. Next press ÷ and enter the first value in the denominator, **52**, press **MATH**, highlight **PRB** and select **3:nCr**, enter the next value, **6.** Press **ENTER** and the answer will be displayed on your screen.

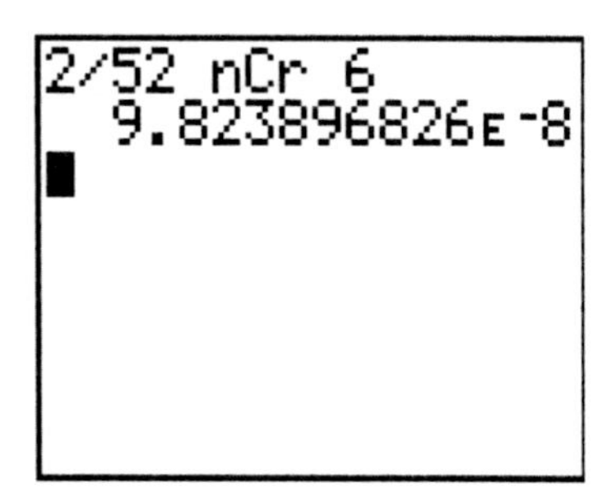

Notice that the answer appears in scientific notation. To convert to standard notation, move the decimal point 8 places to the left. The answer is .0000000982.

Example 15 (pg. 312) Probabilities involving Combinations

In this example, there are 120 fasteners in the shipment. Four fasteners in the shipment are defective. The remaining 116 fasteners are not defective. The quality-control manager randomly selects five fasteners.

To calculate the probability of selecting exactly one defective fastener, you must calculate: $\frac{{}_4C_1 * {}_{116}C_4}{{}_{120}C_5}$

To calculate the numerator, enter the first value, **4**, press **MATH**, highlight **PRB** and select **3:nCr** and enter the next value, **1**. Next press **x** and enter the next value, **116**, press **MATH**, highlight **PRB** and select **3:nCr**, enter the next value, **4.** Next press **÷** and enter the first value in the denominator, **120**, press **MATH**, highlight **PRB** and select **3:nCr**, enter the next value, **5.** Press **ENTER** and the answer will be displayed on your screen.

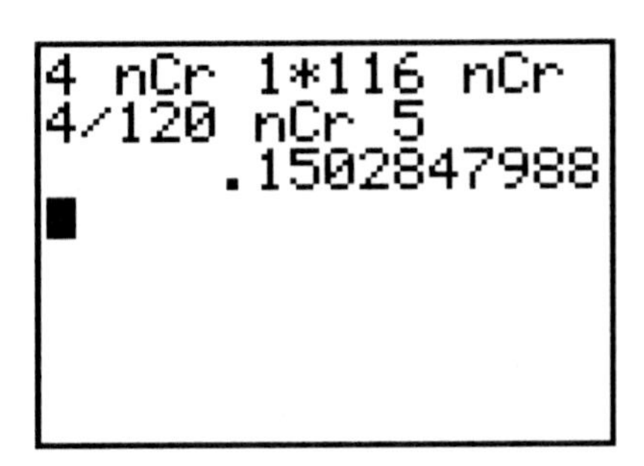

Problem 61 (pg. 314)

In this exercise, there are two groups made up of 8 students and 10 faculty. The combined number in the two groups is 18. Five individuals are to be selected from the total.

(a.) To select all students, you must choose 5 students from the group of 8 students and 0 faculty from the group of 10. Using the combination formula, you will do the following calculation: $\frac{{}_8C_5 * {}_{10}C_0}{{}_{18}C_5}$.

To calculate the numerator, enter the first value, **8**, press **MATH**, highlight **PRB** and select **3:nCr** and enter the next value, **5**. Next press **x** and enter the next value, **10**, press **MATH**, highlight **PRB** and select **3:nCr**, enter the next value, **0.** Next press **÷** and enter the first value in the denominator, **18**, press **MATH**, highlight **PRB** and select **3:nCr**, enter the next value, **5.** Press **ENTER** and the answer will be displayed on your screen.

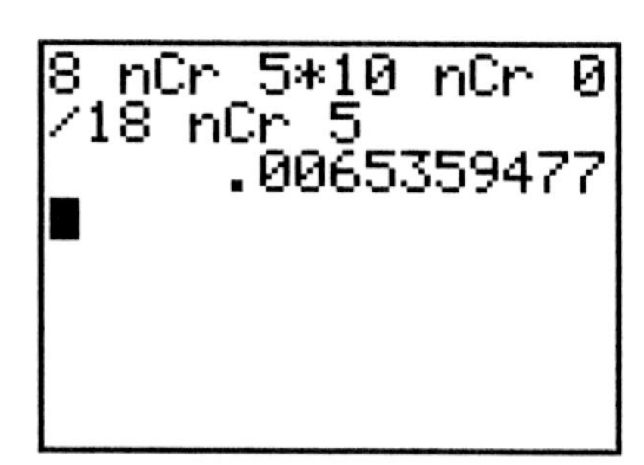

(b.) Repeat the steps in part (a.) but select 0 students from the group of 8 students and 5 faculty members from the group of 10 faculty members.

(c.) Repeat the steps in part (a.) but select 2 students from the group of 8 students and 3 faculty members from the group of 10 faculty members.

‣ Problem 65 (pg. 314)

The compact disk has a total of 13 songs. Of the 13 songs, there are 5 songs that you like and 8 songs that you do not like. Suppose that four songs are randomly selected and played.

(a.) Calculate the probability that, among these first four songs selected, you like exactly two of them. Using the combination formula, you will do the following calculation: $\dfrac{{}_5C_2 * {}_8C_2}{{}_{13}C_4}$.

To calculate the numerator, enter the first value, **5**, press **MATH**, highlight **PRB** and select **3:nCr** and enter the next value, **2**. Next press **x** and enter the next value, **8**, press **MATH**, highlight **PRB** and select **3:nCr**, enter the next value, **2.** Next press **÷** and enter the first value in the denominator, **13**, press **MATH**, highlight **PRB** and select **3:nCr**, enter the next value, **4.** Press **ENTER** and the answer will be displayed on your screen.

(b.) Repeat the steps in part (a.) but select 3 songs from the group of songs that you like and 1 from the group of songs that you do not like.

(c.) Repeat the steps in part (a.) but select 4 songs from the group of songs that you like and 0 songs from the group that you do not like.

◀

TI-89 Instructions:

These instructions are designed to give you an overview of the probability calculations on the TI-89.

The probability menu can be accessed from the **HOME** screen using the keys: **2nd** **MATH** and selecting 7:Probability. Within the probability menu, you well find factorials (!), permutations (nPr) and combinations (nCr).

To calculate a factorial, such as **7!**, go to the data entry line on the **Home** screen. Press **7** **2nd** **MATH**, select **7:Probability** and select **1:!**. Press **ENTER** twice.

To calculate a permutation (or a combination), go to the data entry line on the **Home** screen. Press **2nd** **MATH**, select **7:Probability** and select **2:(nPr)** (or select **3:(nCr)**). Enter the value for **n** (the total number of objects) followed by a comma. Enter the value for **r** (the number of objects being selected). Close the parenthesis and press **ENTER**.

TI-*n*spire Instructions:

These instructions are designed to give you an overview of the probability calculations on the TI-*n*spire handhelds.

Press (home) and select **6:New Document**. (Note: If you currently have a document open, the next screen will ask if you want to save the document. Press (tab) to select **No**. Press (enter).) Select **1:Add Calculator.** To access the probability functions press (menu) and select **5:Probability.**

To calculate a factorial, such as **7!**, type **7,** press (menu), select **5:Probability** and select **1:Factorial (!)**. Press (enter).

To calculate a permutation (or a combination), press (menu), select **5:Probability** and select **2:Permutations (or 3:Combinations).** Enter the value for **n** (the total number of objects) followed by a comma. Enter the value for **r** (the number of objects being selected). Close the parenthesis and press (enter).

Discrete Probability Distributions

Section 6.1

Example 4 (pg. 333) A Probability Histogram

Press **STAT** and select **1:EDIT**. Clear **L1** and **L2**. Using Table 1 from Example 2 on pg. 332, enter the X-values into **L1** and the P(x) values into **L2**.

To graph the probability distribution, first make sure that there is nothing stored in the Y-registers. Press **Y=** and check the Y-registers. If any of them contain a function, move the cursor to that Y-register and press **CLEAR**.

Press **2nd [STAT PLOT]** and press **ENTER**. Turn **ON** Plot 1, select **Histogram** for **Type**, type in **2nd [L1]** for **Xlist** and **2nd [L2]** for **Freq.** Press **WINDOW** and set **Xmin** = 0, **Xmax = 4, Xscl = 1, Ymin = 0** and **Ymax = .55.** Choosing 'Xmax=4' leaves some space at the right of the graph in order to complete the histogram. The Ymax value was selected by looking through the values in **L2** and then rounding the largest value UP to a convenient number. Press **GRAPH** to view the histogram.

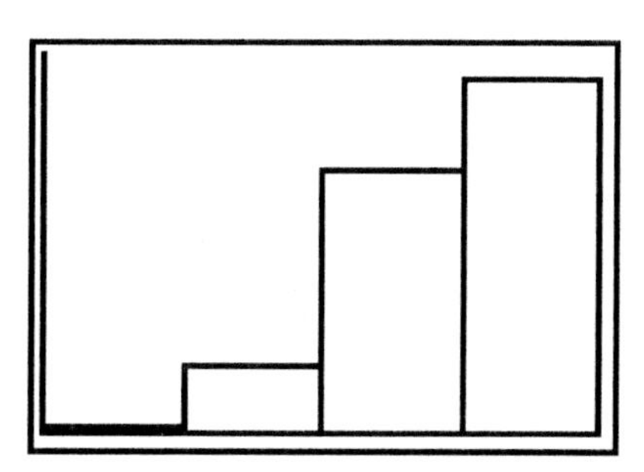

Example 7 (pg. 337) The Expected Value

Press **STAT** and select **1:EDIT**. Clear **L1** and **L2**. Enter the X-values from Table 5 into **L1** and the associated probabilities into **L2**.

```
L1       L2       L3      2
350      .99894   ------
-2.5E5   .00106
------
L2(3) =
```

Notice that the value of –249,650 appears as –2.5E5. This is a rounded value and it is written in scientific notation. The actual value is stored in the calculator; the rounded value is for display purposes only. Also, notice that the values in L2 are displayed as rounded values.

Press **STAT** and highlight **CALC**. Select **1:1-Var Stats,** press **ENTER** and press **2nd** **[L1]** , **2nd** **[L2]** **ENTER** to see the descriptive statistics.

```
1-Var Stats
 x̄=84
 Σx=84
 Σx²=66436300
 Sx=
 σx=8150.413732
↓n=1
```

The expected value of this discrete random variable is $84.00.

▸ Example 9 (pg. 339) The Mean and Standard Deviation of a Discrete Random Variable

Press **STAT** and select **1:EDIT**. Clear **L1** and **L2**. Enter the X-values from Example 2, Table 1 on pg. 332 into **L1** and the P(x) values into **L2**. Press **STAT** and highlight **CALC**. Select **1:1-Var Stats,** press **ENTER** and press **2nd** **[L1]** **,** **2nd** **[L2]** **ENTER** to see the descriptive statistics.

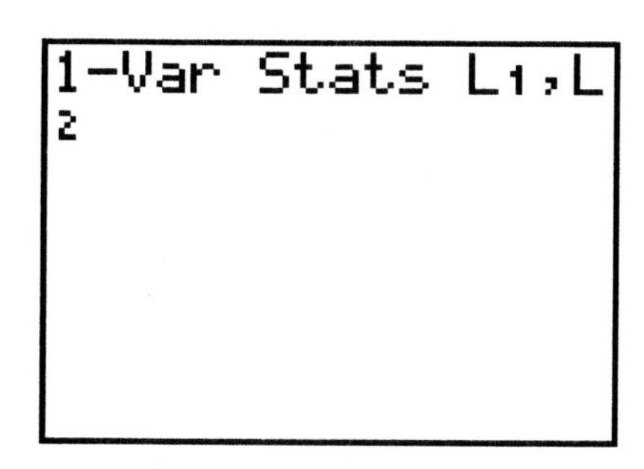


```
1-Var Stats
 x̄=2.39
 Σx=2.39
 Σx²=6.21
 Sx=
 σx=.7056202945
↓n=1
```

The mean of this discrete random variable is 2.39 and the standard deviation is .7056.

▶ Problem 26 (pg. 341)

(a.) Enter the X-values into L1 and the frequencies into L2. Press 2nd [QUIT]. Press 2nd [LIST] and select **MATH**. Select **5:sum(** and type in **L2**. Press **ENTER**. The answer is the sum of the frequencies in L2. Press **STAT** and select **1:EDIT**. Move the cursor to highlight 'L2' at the top of the second list and press **ENTER**. With the cursor flashing at the bottom of the screen, type in **L2** ÷ (the sum of L2). This will convert the frequencies in L2 into probabilities. To confirm that you now have a probability distribution represented in L1 and L2, press 2nd [QUIT]. Press 2nd [LIST] and select **MATH**. Select **5:sum(** and type in **L2**. Press **ENTER**. The answer is the sum of the probabilities in L2. This sum should equal 1.

(b.) To draw the probability histogram, first make sure that there is nothing stored in the Y-registers. Press **Y=** and check the Y-registers. If any of them contain a function, move the cursor to that Y-register and press **CLEAR**. Press **2nd [STAT PLOT]** and press **ENTER**. Turn **ON** Plot 1, select **Histogram** for **Type**, type in **2nd [L1]** for **Xlist** and **2nd [L2]** for **Freq.** Press **WINDOW** and set **Xmin** = 9, **Xmax = 13, Xscl = 1, Ymin = 0** and **Ymax = .26.** Choosing 'Xmax=13' leaves some space at the right of the graph in order to complete the histogram. The Ymax value was selected by looking through the values in **L2** and then rounding the largest value UP to a convenient number. Press **GRAPH** to view the histogram.

(c.-d.) Press **STAT** and highlight **CALC**. Select **1:1-Var Stats,** press **ENTER** and press 2nd [L1] [,] 2nd [L2] **ENTER** to see the descriptive statistics.

◀

Section 6.2

Example 5 (pg. 350) Binomial Probability Distribution Function

(a.) To find the probability that exactly 5 cable television subscribers in a random sample of 15 subscribers owned an HDTV in 2007, we will use the binomial probability density function, **binompdf(n,p,x).** For this example, n = 15, p = .30 and x = 5. Press **2nd [DISTR]**. Scroll down through the menu to select **A:binompdf(** and press **ENTER**. (Note: On the TI-83, the binomial probability distribution function is option **0:binompdf.**) Type in **15** , **.30** , **5**) and press **ENTER**. The answer, **.2061**, will appear on the screen.

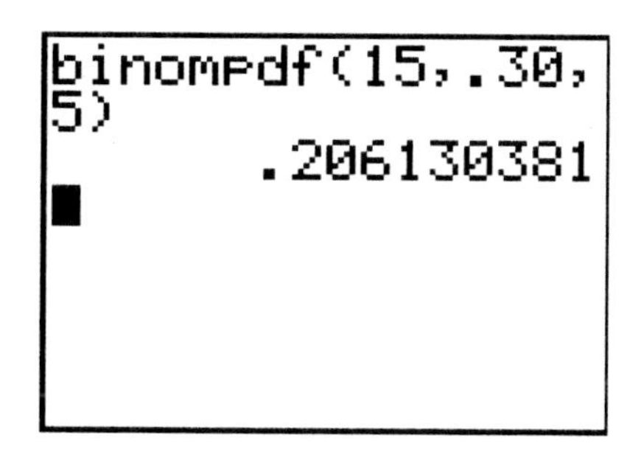

b. To calculate inequalities, such as the probability that *fewer than* 9 subscribers in a group of 15 subscribers owned an HDTV in 2007, that is P(X < 9), you can use the cumulative probability command: **binomcdf (n,p,x)**. This command accumulates probability starting at X = 0 and ending at a specified X-value.

To calculate P(X < 9), we must accumulate probabilities for X = 0, 1, 2, …8, which is P(X ≤ 8). Press **2nd [DISTR]** scroll through the options and select **B:binomcdf(**. Type in **15** , **.30** , **8**) and press **ENTER**. The result, P(X ≤ 8) = P(X < 9) = .9848.

▸ Example 7 (pg. 351) Binomial Probability Histogram

a.) Construct a probability distribution for a binomial probability model with n = 10 and p = 0.2. Press STAT, select **1:EDIT** and clear **L1** and **L2**. Enter the values 0 through 10 into **L1**. Press **2nd [QUIT].**

To calculate the probabilities for each X-value in **L1**, first change the display mode so that the probabilities displayed will be rounded to 3 decimal places. Press MODE and change from **FLOAT** to **3.** Press ENTER. This will round each of the probabilities to 3 decimal places. Press 2nd **[QUIT].**

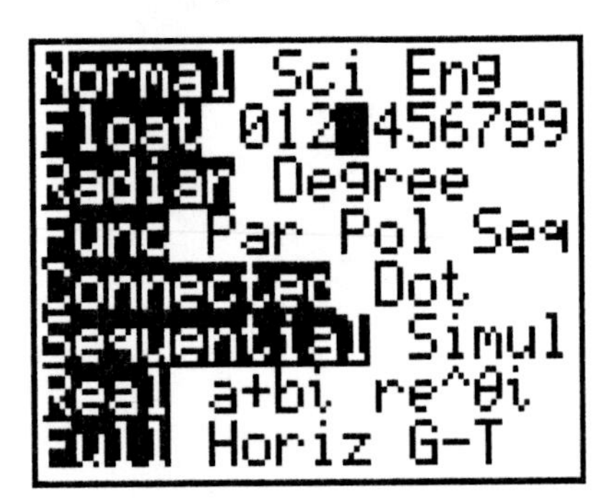

Press STAT, select **1:EDIT.** Move the cursor so that it highlights '**L2**' at the top of **L2**. Press ENTER. Next press **2nd [DISTR]** and select **A:binompdf(** and type in **10** , **.2**) and press ENTER.

L1	L2	L3 3
0.000	.107	
1.000	.268	
2.000	.302	
3.000	.201	
4.000	.088	
5.000	.026	
6.000	.006	

L3(1)=

To graph the binomial distribution, first make sure that there is nothing stored in the Y-registers. Press Y= and check the Y-registers. If any of them contain a function, move the cursor to that Y-register and press CLEAR.
Press **2nd [STAT PLOT]** and press ENTER. Turn **ON** Plot 1, select **Histogram** for **Type**, type in **2nd [L1]** for **Xlist** and **2nd [L2]** for **Freq.** Adjust the graph window by pressing WINDOW and setting **Xmin** = 0, **Xmax = 11, Xscl = 1, Ymin = 0** and **Ymax = .31.** Choosing 'Xmax=11' leaves some space at the right of the graph in order to complete the histogram. The Ymax value was selected by looking through the values in **L2** and then rounding the largest value UP to a convenient number. Press GRAPH to view the histogram.

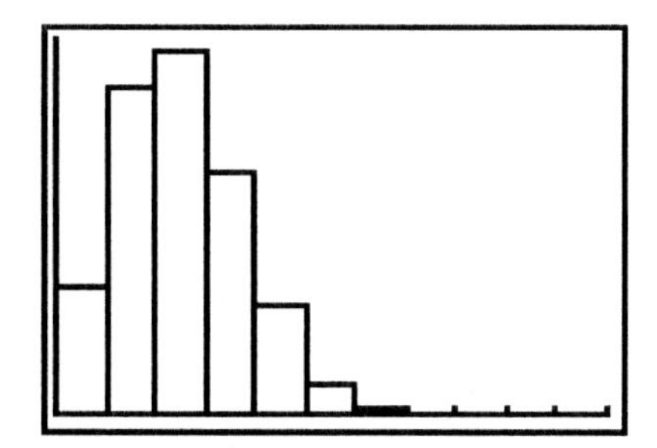

◀

▸ Problem 31 (pg. 354)

(a.) Construct a probability distribution for a binomial probability distribution with n = 9 and p = .75. Press **STAT**, select **1:EDIT** and clear **L1** and **L2**. Enter the values 0 through 9 into **L1**. Press **2nd [QUIT]**. To calculate the probabilities for each X-value in **L1**, first change the display mode so that the probabilities displayed will be rounded to 3 decimal places. Press **MODE** and change from **FLOAT** to **3** and press **2nd [QUIT]**.

Press **STAT**, select **1:EDIT.** Move the cursor so that it highlights '**L2**' at the top of **L2**. Press **ENTER**. Next press **2nd [DISTR]** and select **A:binompdf(** and type in **9** [,] **.75** [)] and press **ENTER**.

(b.) Press **STAT** and highlight **CALC**. Select **1:1-Var Stats,** press **ENTER** and press **2nd [L1]** [,] **2nd [L2]** **ENTER** to see the descriptive statistics. The mean (μ) of the random variable is displayed as $\bar{x}$ on the calculator. The standard deviation is σx.

(c.) Use the formulas for the mean and standard deviation of a binomial random variable. The mean is: $\mu = n * p$; the standard deviation is: $\sigma = \sqrt{n * p * q}$.

(d.) To draw the probability histogram, first make sure that there is nothing stored in the Y-registers. Press **Y=** and check the Y-registers. If any of them contain a function, move the cursor to that Y-register and press **CLEAR**. Press **2nd [STAT PLOT]** and press **ENTER**. Turn **ON** Plot 1, select **Histogram** for **Type**, type in 2nd [L1] for **Xlist** and 2nd [L2] for **Freq.** Adjust the graph window by pressing **WINDOW** and setting **Xmin** = 0, **Xmax = 10, Xscl = 1, Ymin = 0** and **Ymax = .31.** Choosing 'Xmax=10' leaves some space at the right of the graph in order to complete the histogram. The Ymax value was selected by looking through the values in **L2** and then rounding the largest value UP to a convenient number. Press **GRAPH** to view the histogram.

◂

▸ Problem 55 (pg. 357)

(a.) To generate random samples for this binomial model, press **MATH**, select **PRB** and select **7:randBin(**. This command requires three values: **n**, which is the sample size; **p**, the probability; and **x**, the number of samples. For this example type in **30** [,] **.98** [,] **100).** Press **ENTER** It will take the calculator a few minutes to complete this simulation.

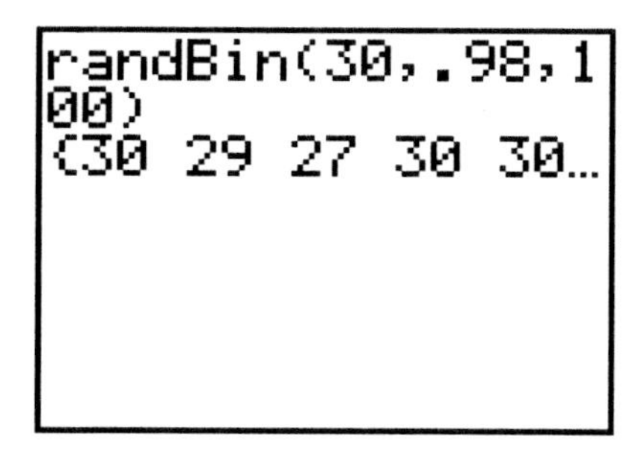

Store these probabilities in **L1** by pressing **STO** **2nd** **[L1]** .

(b.) To use the results of the simulation to compute the probability that exactly 29 of the 30 males survive to age 30, construct a histogram of the simulation. First make sure that there is nothing stored in the Y-registers. Press **Y=** and check the Y-registers. If any of them contain a function, move the cursor to that Y-register and press **CLEAR**.
Press **2nd** **[STAT PLOT]** and press **ENTER**. Turn **ON** Plot 1, select **Histogram** for **Type**, type in **2nd** **[L1]** for **Xlist** and **1** for **Freq.** Adjust the graph window by pressing **WINDOW** and setting **Xmin = 26**, **Xmax = 31, Xscl = 1, Ymin = 0** and **Ymax = 60.** (Note: Choosing these values for Xmin and Xmax will display a small part of the probability histogram centered around the value of 29.) Press **GRAPH** to view the partial histogram. Press **TRACE** and scroll through the bars until you reach the bar for '29'. Take the frequency for that bar and divide it by 100 (the total number of simulations). Your result is the probability that exactly 29 males in a sample of 30 males will survive to age 30.

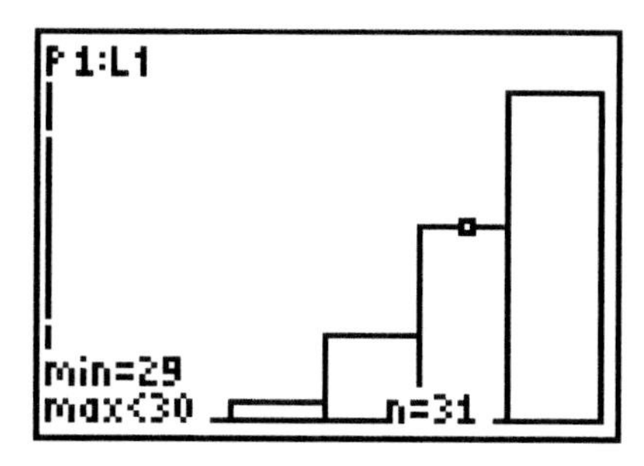

In this simulation, the probability is 31 out of 100 or 31%.

(c.) Press **2nd DISTR** and select **A:binompdf** and enter **30 , .98 , 29).**

(d.) Press GRAPH and the partial histogram of the simulation will appear. Press TRACE and scroll through the bars for '28', '29' and '30'. Sum the frequencies for these bars. Divide this sum by 100. This value is P(X≥28). The *complement* of this is P(X≤ 27). Subtract P(X≥28) from 1 to get P(X≤ 27).

(e.) Press **2nd DISTR** and select **B:binomcdf** and enter **30 , .98 , 27).** This value is P(X≤ 27).

(f-g.) First, calculate the mean and standard deviation of the 100 simulations. Press STAT and highlight **CALC**. Select **1:1-Var Stats,** press ENTER and press 2nd [L1] ENTER to see the descriptive statistics. The mean (μ) of the random variable is displayed as $\bar{x}$ on the calculator. The standard deviation is σx.

Next, use the formulas for the mean and standard deviation of a binomial random variable. The mean is: $\mu = n * p$; the standard deviation is: $\sigma = \sqrt{n * p * q}$.

◀

Problem 58 (pg. 357)

(a.) Suppose the probability that Shaquille O'Neal makes a *free throw* is .524. To find the probability that the *first* free throw he makes occurs on his third shot, press **2nd [DISTR]** and select **E:geometpdf(** and type in **.524 , 3)**.

```
geometpdf(.524,3
)
        .118725824
```

(b.) Construct a probability distribution for a geometric probability model with p = .524. Press **STAT**, select **1:EDIT** and clear **L1** and **L2**. Enter the values 1 through 10 into **L1**. Press **2nd [QUIT]**. To calculate the probabilities for each X-value in **L1**, first change the display mode so that the probabilities displayed will be rounded to 3 decimal places. Press **MODE** and change from **FLOAT** to **3** and press **2nd [QUIT]**.

Press **STAT**, select **1:EDIT.** Move the cursor so that it highlights '**L2**' at the top of **L2**. Press **ENTER**. Next press **2nd [DISTR]** and select **E:geometpdf(** and type in **.524 , L1)** and press **ENTER**.

(c.) Press **STAT** and highlight **CALC**. Select **1:1-Var Stats,** press **ENTER** and press 2nd [L1] , 2nd [L2] **ENTER** to see the descriptive statistics. The mean (μ) of the random variable is displayed as $\bar{x}$ on the calculator.

(d.) Calculate the mean of a geometric probability model using the formula: $\mu = \frac{1}{p}$.

Section 6.3

▸ Example 2 (pg. 360) Probabilities of a Poisson Process

For problems that can be modeled with the poisson probability model, either the values of λ and **t** are given or the value of μ is given. These parameters (**λ, t and** μ) are related to each other in the following way: **μ=λ*t.** The **poissonpdf** command requires a value for **μ.**

(a.) In this example, **λ= 2** (cars per minute) and **t= 5** (minutes) and, therefore, **μ=2*5=10.** Use the command **poissonpdf (μ,x)** with μ = 10 and X = 6. Press 2nd [DISTR] and select **C:poissonpdf(** and type in **10** , **6**) and press ENTER. The answer will appear on the screen.

```
poissonpdf(10,6)
          .063055458
```

(b.) To calculate the probability that *less than 6* cars arrive in the 5 minute time period, use the command **poissoncdf (μ,x)** with μ = 10 and X = 5. Press **2nd [DISTR]** and select **D:poissoncdf(** and type in **10** , **5**) and press ENTER. The answer will appear on the screen.

```
poissoncdf(10,5)
         .0670859629
```

(c.) P(X≥ 6) = 1 – P(X≤5) = 1 - .0671 = .9329.

Problem 7 (pg. 362)

This is an example of a Poisson process with **μ=5.**

(a.) Press **2nd [DISTR]** and select **C:poissonpdf(** and type in 5 , 6).

(b.) To calculate P(X < 6) press **2nd [DISTR]** and select **D:poissoncdf(** and type in 5 , 5).

(c.) P(X ≥ 6) = 1 – P(X < 6)

(d.) Press **2nd [DISTR]** and select **C:poissonpdf(** and type in 5 , **2nd** { 2 , **3** , **4** **2nd** }) and press **ENTER**.

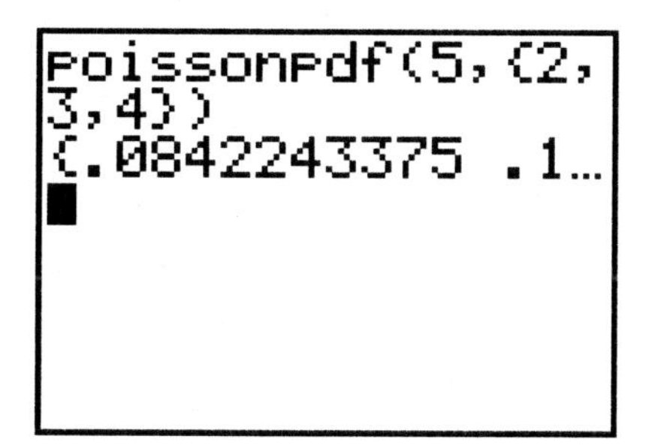

The first answer that appears in the output is P(2) which is .0842. Use the right arrow to scroll to the right to see P(3) and P(4).

TI-89 Instructions:

These instructions are designed to give you an overview of the discrete probability distribution functions on the TI-89.

Select the **Stat/List** icon. The screen that follows allows you to store the data in specific folders. It is easy to always use the **Main** folder so simply press ENTER. The next screen displays Lists 1 through 4. If any Lists contain data and you want to delete that data, press **F3**, select **1:Names** and scroll down to each List that you would like to delete. While the list is highlighted, press **F4** to select it. Once you have selected all the lists that you want to delete, press **F1** and select **1:Delete** and press ENTER. Press ESC to return to the lists. Notice that the lists that you just deleted are no longer displayed. Press **F1** and select **3:Setup Editor** and press ENTER. Lists 1 through 4 are again displayed.

Calculating μ and σ for a discrete random variable:
To calculate μ and σ for a discrete random variable, enter the x-values into **List1** and enter the probabilities into **List2**. Press **F4** and choose **1:OneVar**. Use 2nd VAR-LINK to select **list1** for **List** and **list2** for **Freq**. Press ENTER to display the statistics.

Accessing the Probability Distributions:
To access the probability functions (i.e., binomial, normal, poisson, geometric, etc.) press **F5** and a menu of functions will appear. Suppose you want the binomial probability density function. Scroll to B:Binomial pdf and press ENTER. Enter the **number of trials, the probability of success** and the **x-value**. Press ENTER.

TI-*n*spire Instructions:

These instructions are designed to give you an overview of the discrete probability distribution functions on the **TI-*n*spire** handhelds.

Calculating μ and σ for a discrete random variable:
Press (home) and select **6:New Document**. (Note: If you currently have a document open, the next screen will ask if you want to save the document. Press (tab) to select **No**. Press (enter).) Select **3:Add Lists & Spreadsheet.** Enter the x-values into **column A** and enter the probabilities into **column B**. Press (menu), select **4:Statistics** and select **1: Stat Calculations.** Select **1:One-Variable Statistics**. On the next screen, set **Num of Lists** to 1. Press (tab) to select **OK** and press (enter). On the next screen, in the first entry box (**X list**) type **a** (for Column A). Press (tab) to scroll to the next line (**Frequency**) and type **b** (for Column B). Press (tab) to scroll to the 1st Result Column and type **c** (for Column C). Press (tab) to select **OK**. Press (enter). The spreadsheet contains two new columns, Column C and Column D. Column C contains the labels of the numerical values that are displayed in Column D. Use the arrow keys (▲ ▶ ◀ ▼) to move to the top of column D and press (menu). Select **1:Actions, 2:Resize** and **1:Resize Column Width.** Click the right arrow until Column D has been expanded to a larger width. Press the down arrow to remove the shading. Use the down arrow to scroll down through the values in Column D.

Accessing the Probability Distributions:
Press (home) and select **6:New Document**. (Note: If you currently have a document open, the next screen will ask if you want to save the document. Press (tab) to select **No**. Press (enter).) Press (enter) to select **1:Add Calculator.** To access the probability functions (i.e., binomial, normal, poisson, geometric, etc.), press (menu), select **5:Probability** and select **5:Distributions**. A menu of probability distributions will be displayed in a list on the left side of the screen. For example, suppose you have a situation that you would model with the binomial probability density function. Scroll to **D:Binomial pdf** and press (enter). Enter the **number of trials**, press (tab) to scroll to the next line, enter **the probability of success,** press (tab) to scroll to the next line and enter the **x-value**. Press (tab) to highlight **OK** and press (enter).

The Normal Probability Distribution

Section 7.2

▸ Example 2 (pg. 386) Area Under the Standard Normal Curve to the left of a Z-score

In this example of the standard normal curve, we will calculate the area to the *left* of Z = 1.68.

The TI-84 has two methods for calculating this area.

Method 1: **Normalcdf**(lowerbound, upperbound, μ, σ) computes the area between a lowerbound and an upperbound. In this example, you are computing the area from *negative infinity* to 1.68. Negative infinity is specified by (-) 1 2nd **[EE]** 9 9 (Note: **EE** is found above the comma ,). Try entering –1 EE 99 into your calculator.

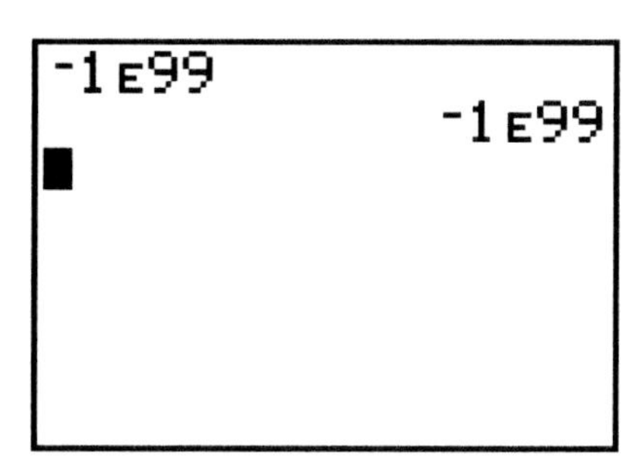

Now, to calculate the area to the left of 1.68, press **2nd** **[DISTR]** and select **2:normalcdf(** and type in **-1E99** , **1.68** , **0** , **1**) and press **ENTER**. (Note: For the standard normal curve, μ = **0** and σ **=1.**)

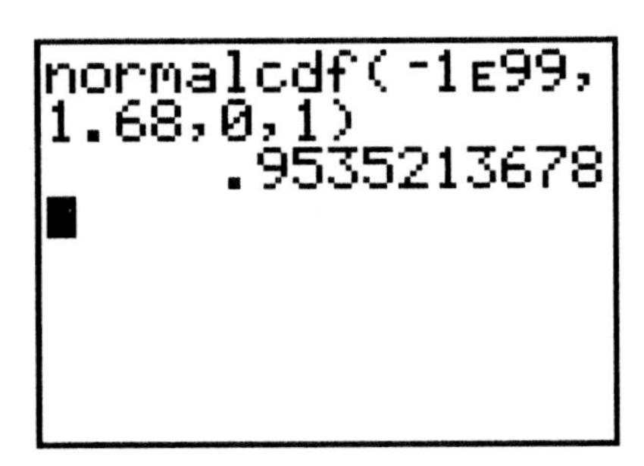

Method 2: This method calculates the area and also displays a graph of the probability distribution. You must first set up the WINDOW so that the graph will be displayed properly. Press WINDOW and set **Xmin** equal to -3 and set **Xmax** equal to 3. Set **Xscl** equal to 1.

Setting the Y-range is a little more difficult to do. A good "rule - of - thumb" is to set **Ymax** equal to .5 / σ . For this example, set **Ymax = .5.**

Use the up arrow to highlight **Ymin**. A good value for **Ymin** is **(-) Ymax / 4** so type in (-) **.5 / 4**.

Press 2nd [QUIT]. Clear all the previous drawings by pressing 2nd **[DRAW]** and selecting **1:ClrDraw** and pressing ENTER ENTER. Press 2nd **[STATPLOT]** and TURN OFF all PLOTS. Make sure that there is nothing stored in the Y-registers. Press Y= and check the Y-registers. If any of them contain a function, move the cursor to that Y-register and press CLEAR. Now you can draw the probability distribution. Press **2nd [DISTR]**. Highlight **DRAW** and select **1:ShadeNorm(** and type in **-1E99** , **1.68** , **0** , **1**) and press ENTER. The output displays a normal curve with the appropriate area shaded in and its value computed.

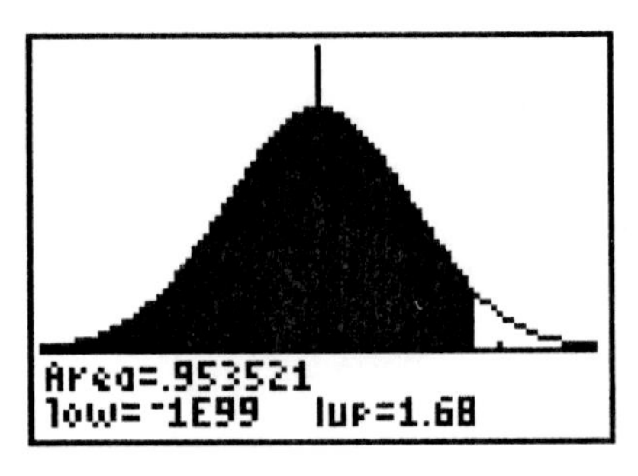

.5909

▸ Example 3 (pg. 387) Area Under the Standard Normal Curve to the right of a Z-score

In this example of the standard normal curve, we will calculate the area to the *right* of Z = -0.46.

Method 1: **Normalcdf**(lowerbound, upperbound, μ, σ) computes the area between a lowerbound and an upperbound. In this example, you are computing the area from –0.46 to *positive infinity*. Positive infinity is specified by 1 2nd **[EE]** 9 9 (Note: **EE** is found above the comma ,). To calculate the area to the right of –0.46, press **2nd [DISTR]** and select **2:normalcdf(** and type in **–0.46** , **1E99** , **0** , **1**) and press ENTER.

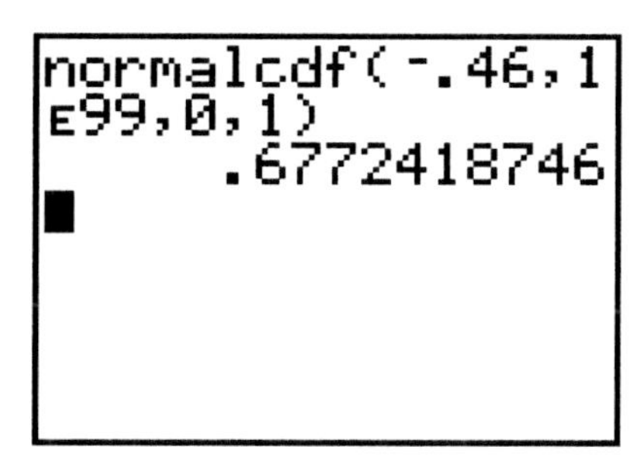

Method 2: This method calculates the area and also displays a graph of the probability distribution. Press WINDOW and set **Xmin** equal to -3 and set **Xmax** equal to 3. Set **Xscl** equal to 1. Set **Ymax = .5.** Set **Ymin = .5/4.** Press 2nd [QUIT]. Clear all the previous drawings by pressing 2nd **[DRAW]** and selecting **1:ClrDraw** and pressing ENTER ENTER. Press 2nd **[STATPLOT]** and **TURN OFF** all **PLOTS**. Make sure that there is nothing stored in the Y-registers. Press Y= and check the Y-registers. If any of them contain a function, move the cursor to that Y-register and press CLEAR. Now you can draw the probability distribution. Press **2nd [DISTR]**. Highlight **DRAW** and select **1:ShadeNorm(** and type in **–0.46** , **1E99** , **0** , **1**) and press ENTER.

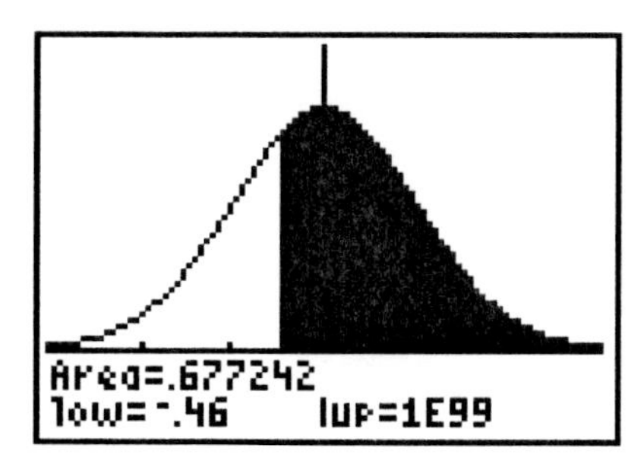

◀

▸ Example 4 (pg. 388) Area Under the Standard Normal Curve between two Z-scores

In this example of the standard normal curve, we will calculate the area between Z= -1.35 and Z= 2.01.

Method 1: **Normalcdf**(lowerbound, upperbound, μ, σ) computes the area between a lowerbound and an upperbound. In this example, you are computing the area from –1.35 to 2.01. To calculate the area between –1.35 and 2.01, press 2nd **[DISTR]** and select **2:normalcdf(** and type in **–1.35** , **2.01** , **0** , **1**) and press **ENTER**.

```
normalcdf(-1.35,
2.01,0,1)
          .8892764236
```

Method 2: This method calculates the area, and also displays a graph of the probability distribution. Press **WINDOW** and set **Xmin** equal to -3 and set **Xmax** equal to 3. Set **Xscl** equal to 1. Set **Ymax = .5.** Set **Ymin = .5/4.**

Press 2nd [QUIT]. Clear all the previous drawings by pressing 2nd **[DRAW]** and selecting **1:ClrDraw** and pressing **ENTER** **ENTER**. Press 2nd **[STATPLOT]** and **TURN OFF** all **PLOTS**. Make sure that there is nothing stored in the Y-registers. Press **Y=** and check the Y-registers. If any of them contain a function, move the cursor to that Y-register and press **CLEAR**. Now you can draw the probability distribution. Press 2nd [DISTR]. Highlight **DRAW** and select **1:ShadeNorm(** and type in **–1.35** , **2.01** , **0** , **1**) and press **ENTER**.

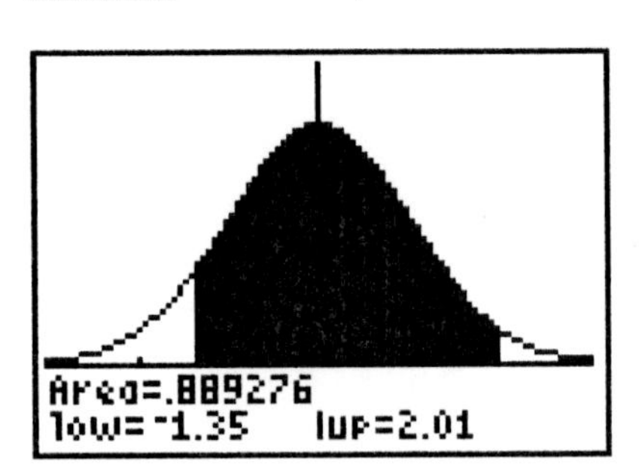

▸ Example 6 (pg. 390) Finding a Z-Score from a Specified Area to the Left

This is called an inverse normal problem and the command **invNorm(area,** μ **,** σ **)** is used. In this type of problem, an area under the normal curve is given and you are asked to find the corresponding Z-score. In this example, the area given is the area to the *left* of a Z-score. The area is 0.32 . (The area value that you **ENTER** into the TI-84must always be area to the left of a Z-score.) .

To find the Z-score corresponding to *left area* of 0.32, press **2nd [DISTR]** and select **3:invNorm(** and type in **.32** [,] **0** [,] **1**) and press **ENTER**.

```
invNorm(.32,0,1)
      -.4676988012
█
```

The Z-score of -.47 has an area of 0.32 to the *left*.

◀

▸ Example 7 (pg. 391) Finding a Z-Score from a Specified Area to the Right

This is an inverse normal problem and the command **invNorm(area,** μ **,** σ **)** is used. In this type of problem, an area under the normal curve is given and you are asked to find the corresponding Z-score. In this example, the area given is the area to the *right* of a Z-score. The area is 0.4332 . (The area value that you enter into the TI-84 must always be area to the left of a Z-score.) .

To find the Z-score corresponding to *right area* of 0.4332, subtract 0.4332 from 1 to obtain the area to the *left* of the Z-score. Press **2nd [DISTR]** and select **3:invNorm(** and type in **.5668** [,] **0** [,] **1**) and press ENTER.

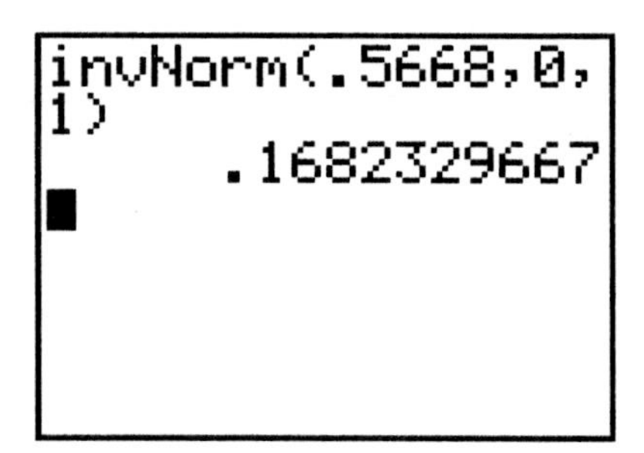

The Z-score that has an area to the *right* equal to 0.4332 is Z= .17.

▶ Example 8 (pg. 391) Finding Z-Scores for an Area in the Middle

This is an inverse normal problem and the command **invNorm(area,** μ **,** σ **)** is used. In this problem the *middle area* is .90. That leaves an area of .10 to be equally divided between the *left* and *right* tail areas. Each of these areas are, therefore, equal to .05. The Z-score that marks the lower edge of the middle area is the Z-score that corresponds to a *left area* of .05.

To find the Z-score corresponding to a *left area* of 0.05, press **2nd [DISTR]** and select **3:invNorm(** and type in **.05** [,] **0** [,] **1**) and press ENTER.

```
invNorm(.05,0,1)
    -1.644853626
```

The Z-score that marks the lower edge of the middle area of .90 is **–1.645.**

To find the Z-score corresponding to a *left area* of 0.95, press 2nd [DISTR] and select **3:invNorm(** and type in **.95** [,] **0** [,] **1**) and press ENTER. The Z-score that marks the upper edge is **1.645.**

◀

Section 7.3

▸ Example 2 (pg. 398) Finding Area Under a Normal Curve

In this exercise, use a normal distribution with μ = 38.72 and σ = 3.17.

Method 1:To find the percentile rank of a three-year-old female whose height is 43 inches, we calculate P(X < 43). Press **2nd** **[DISTR]**, select **2:normalcdf(** and type in **-1E99** , **43** , **38.72** , **3.17**) and press ENTER.

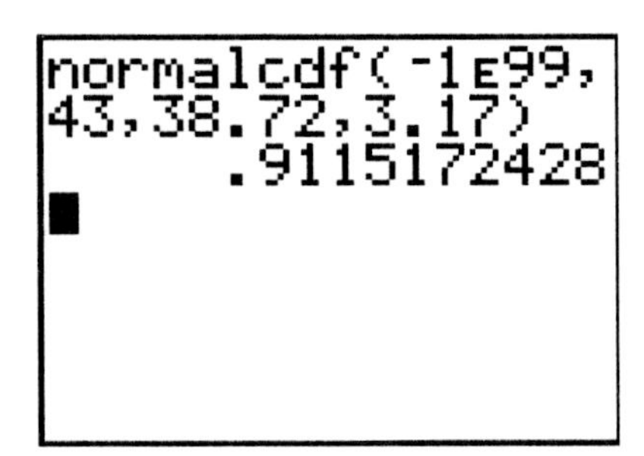

Method 2: To find P(X < 43) and include a graph, you must first set up the WINDOW so that the graph will be displayed properly. You will need to set Xmin equal to (μ - 3 σ) and Xmax equal to (μ + 3 σ). Press WINDOW and set **Xmin** equal to (μ - 3 σ) by typing in **38.72 - 3 * 3.17.** Press ENTER and set **Xmax** equal to (μ + 3 σ) by typing in **38.72 + 3 * 3.17.** Set **Xscl** equal to σ, which is 3.17.

Setting the Y-range is a little more difficult to do. A good "rule - of - thumb" is to set **Ymax** equal to .5 / σ. For this example, set **Ymax = .5/3.17.**

Use the up arrow to highlight **Ymin**. A good value for **Ymin** is **(-) Ymax / 4** so type in (-) **.158 / 4**.

Press **2nd** **[DRAW]** and select **1:ClrDraw** and press ENTER ENTER. Press 2nd **[STATPLOT]** and **TURN OFF** all **PLOTS**. Make sure that there is nothing stored in the Y-registers. Press Y= and check the Y-registers. If any of them contain a function, move the cursor to that Y-register and press CLEAR. Press **2nd** **[DISTR]**, highlight **DRAW** and select **1:ShadeNorm(** and type in **-1E99** , **43** , **38.72** , **3.17**) and press ENTER.

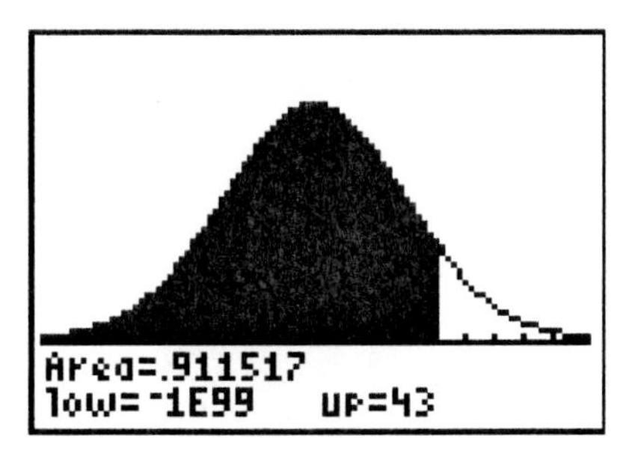

Conclusion: 91 % of all three-year-old females are less than 43 inches tall.

Note: When using the TI-84 (or any other technology tool), the answers you obtain may vary slightly from the answers that you would obtain using the standard normal table. Consequently, your answers may not be exactly the same as the answers found in your textbook. The differences are simply due to rounding.

◀

▸ Example 3 (pg. 399) Finding the Probability of a Normal Random Variable

In this exercise, use a normal distribution with μ = 38.72 and σ = 3.17.

Method 1:To find P(35 ≤ X ≤ 40) press **2nd [DISTR]**, select **2:normalcdf(** and type in **35** , **40** , **38.72** , **3.17**) and press **ENTER**.

```
normalcdf(35,40,
38.72,3.17)
      .5365173032
```

Method 2: To find the probability and include a graph, you must first set up the WINDOW so that the graph will be displayed properly. You will need to set Xmin equal to (μ - 3 σ) and Xmax equal to (μ + 3 σ). Press **WINDOW** and set **Xmin** equal to (μ - 3 σ) by typing in **38.72 - 3 * 3.17.** Press **ENTER** and set **Xmax** equal to (μ + 3 σ) by typing in **38.72 + 3 * 3.17.** Set **Xscl** equal to σ, which is 3.17.

Setting the Y-range is a little more difficult to do. A good "rule - of - thumb" is to set **Ymax** equal to .5 / σ. For this example, set **Ymax = .5/3.17.**

Use the up arrow to highlight **Ymin**. A good value for **Ymin** is **(-) Ymax / 4** so type in (-) **.158 / 4**.

Press 2nd [DRAW] and select 1:ClrDraw and press ENTER ENTER. Press 2nd **[STATPLOT]** and **TURN OFF** all **PLOTS**. Make sure that there is nothing stored in the Y-registers. Press **Y=** and check the Y-registers. If any of them contain a function, move the cursor to that Y-register and press **CLEAR**. Press **2nd [DISTR]**, highlight **DRAW** and select **1:ShadeNorm(** and type in **35** , **40** , **38.72** , **3.17**) and press **ENTER**.

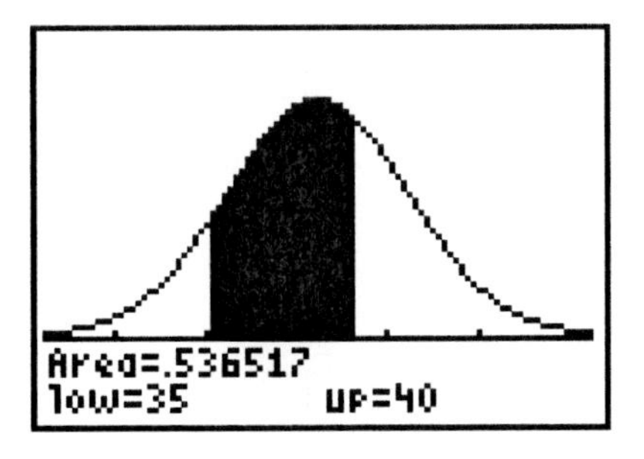

Conclusion: The probability that a randomly selected three-year-old female is between 35 and 40 inches tall is .5365 or 53.65%.

◀

▸ Example 5 (pg. 400) Finding the Value of a Normal Random Variable

This is an inverse normal problem and the command **invNorm(area, μ, σ)** is used. In this type of problem, a percentage of the area under the normal curve is given and you are asked to find the corresponding X-value. In this example, the percentage given is the bottom 20 %, (the 20th percentile). Press **2nd [DISTR]** and select **3:invNorm(** and type in **.20** , **38.72** , **3.17**) and press **ENTER**.

```
invNorm(.20,38.7
2,3.17)
      36.05206069
```

Conclusion: The height that separates the bottom 20% of three-year-old females from the top 80% is 36.05 inches.

◂

▸ Example 6 (pg.401) Finding the Value of a Normal Random Variable

This is an inverse normal problem and the command **invNorm(area,** μ **,** σ **)** is used. In this problem the *middle area* is .98. That leaves an area of .02 to be equally divided between the *left* and *right* tail areas. Each of these areas are, therefore, equal to .01.

The height that marks the lower edge of the middle 98% corresponds to a left area of .01. To find the X-value corresponding to a *left area* of 0.01, press 2nd [DISTR] and select **3:invNorm(** and type in **.01** [,] **38.72** [,] **3.17**) and press **ENTER**.

The height that separates the middle 98% from the top 1% is actually the height that separates the bottom 99% (the middle 98% plus the 1% in the left tail) from the top 1%. To find the X-value corresponding to a *left area* of 0.99, press 2nd [DISTR] and select **3:invNorm(** and type in **.99** [,] **38.72** [,] **3.17**) and press **ENTER**.

```
invNorm(.01,38.7
2,3.17)
      31.34547723
invNorm(.99,38.7
2,3.17)
      46.09452277
```

The middle 98% of the distribution lies between 31.34 inches and 46.09 inches.

◀

Section 7.4

▸ Example 2 (pg. 407) A Normal Probability Plot

Press **STAT** and select **1:Edit** and press **ENTER**. Clear all data from **L1.** **ENTER** the data from Table 4 on pg. 406 into **L1**.

To set up the normal probability plot, first make sure that there is nothing stored in the Y-registers. Press **Y=** and check the Y-registers. If any of them contain a function, move the cursor to that Y-register and press **CLEAR**.
Press **2nd [STAT PLOT]**. Press **ENTER** to select **Plot 1**. Highlight **On** and press **ENTER**. Set **Type** to the normal probability plot which is the third selection in the second row. Press **ENTER**. Set **Data List** to **L1** and **Data Axis** to **X**. Next, there are three different types of **Marks** that you can select for the graph. The first choice, a small square, is the best one to use.

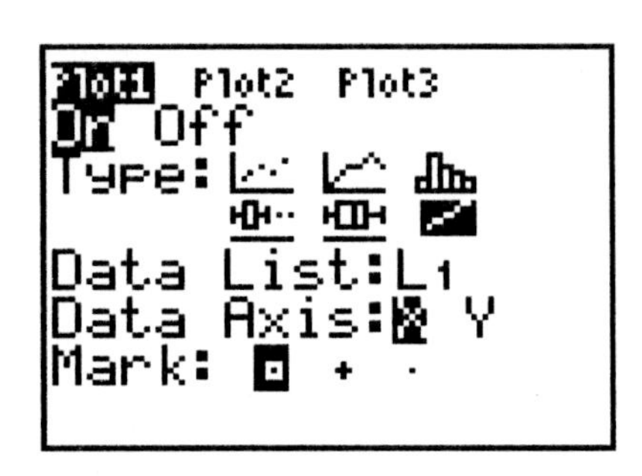

Press **ZOOM** and select **9:ZoomStat** and **ENTER**.

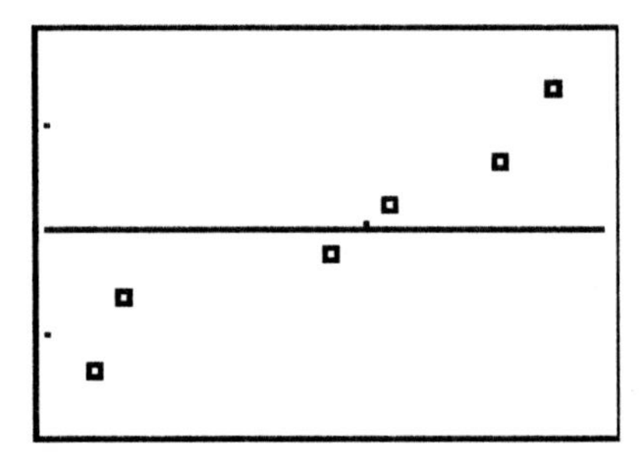

The calculator draws the normal probability plot with a horizontal line at the X-axis. This plot is *fairly* linear, indicating that the data generally follows a normal distribution.

◂

Section 7.5

Example 1 (pg. 414) Normal Approximation to the Binomial

In this binomial experiment, the random variable, X, is the number of individuals with blood type O-negative, the probability, p, that an individual has type O-negative blood is .06 and the sample size, n, is 500. We will approximate the probability that fewer than 25 individuals in the sample have type O-negative blood, that is, $P(X < 25)$ using the normal approximation to the binomial. In order to use the normal approximation to the binomial, the first step is to verify that $n*p*(1-p) \geq 10$. In this example, 500 * .06 * .94 = 28.2, so the requirement is satisfied.

Next, calculate the mean and standard deviation of this binomial random variable. The mean, μ, equals n*p = 500 * .06 = 30. The standard deviation, σ, equals

$$\sqrt{n*p*(1-p)} = \sqrt{500*.06*.94} = 5.31.$$

To approximate $P(X < 25)$ with a normal probability we calculate $P(X \leq 24.5)$. (Note: This adjustment from 25 to 24.5 is called a *continuity correction*).

Press **2nd** **[DISTR]**, select **2:normalcdf(** and type in **-1E99** , **24.5** , **30** , **5.31**) and press **ENTER**.

```
normalcdf(-1E99,
24.5,30,5.31)
        .150152054
```

You can compare this probability (.1505) that you obtained through a normal approximation to the actual probability obtained from the binomial distribution.

Press **2nd** **[DISTR]**, select **B:binomcdf(** and type in **500** , **.06** , **24**) and press **ENTER**.

```
binomcdf(500,.06
,24)
        .1493809338
```

The actual P(X < 25) is .1494.

▸ Example 2 (pg. 415) Normal Approximation to the Binomial

In this binomial experiment, the random variable, X, is the number of households with cable TV, the probability, p, that a household has cable TV is .70 and the sample size, n, is 1000. We will approximate the probability that at least 734 households in the sample have cable TV, that is, $P(X \geq 734)$ using the normal approximation to the binomial. In order to use the normal approximation to the binomial, the first step is to verify that $n*p*(1-p) \geq 10$. In this example, 1000 * .70 * .30 = 210, so the requirement is satisfied.

First, calculate the mean and standard deviation of this binomial random variable. The mean, μ, equals n*p = 1000 * .70 = 700. The standard deviation, σ, equals

$$\sqrt{n*p*(1-p)} = \sqrt{1000*.70*.30} = 14.491.$$

To approximate $P(X \geq 734)$ with a normal probability we calculate $P(X \geq 733.5)$ (Note: This adjustment from 734 to 733.5 is called a *continuity correction*).

Press 2nd [DISTR], select **2:normalcdf(** and type in **733.5** [,] **1E99** [,] **700** [,] **14.491** [)] and press [ENTER].

```
normalcdf(733.5,
1E99,700,14.491)
      .0103948862
```

The approximate $P(X \geq 734)$ is .0104, which is small probability. This suggests that the percentage of households in DuPage County with cable TV is actually higher than 70%.

◂

TI-89 Instructions:

These instructions are designed to give you an overview of the normal probability distribution functions on the TI-89.

The TI-89 has two methods for finding areas under the normal curve and displaying the probabilities.

Method 1: The area under the normal curve is calculated and a graph of the curve is displayed. To use this method, begin by turning off all other graphs. First check the y-registers by pressing ◆ and **Y=** and clearing all the y-functions. In the Stats/List Editor, press **F2** and select **1:Plot Setup**. Press **F3** to clear all plots. Press **ESC** to return to the Editor. Press **F5** and select **1:Shade** and select **1:Shade Normal**. Fill in the entry boxes in the **Shade Normal** screen (lower bound, upper bound, μ and σ.) Press **ENTER**.

Method 2: The probability is calculated without a display of the normal curve. In the Stats/List Editor, press **F5** and select **4:Normal Cdf.** Fill in the entry boxes in the **Normal Cdf** screen (lower bound, upper bound, μ and σ.) Press **ENTER**.

To calculate an x-value that corresponds to a given left-sided area under the normal curve, press **F5** and select **2:Inverse** and select **1:Inverse Normal.** Fill in the entry boxes (left-sided area, μ and σ). Press **ENTER**.

TI-*n*spire Instructions:

These instructions are designed to give you an overview of the normal probability distribution functions on the TI-*n*spire handhelds.

Press (home) and select **6:New Document**. (Note: If you currently have a document open, the next screen will ask if you want to save the document. Press (tab) to select **No**. Press (enter).) Select **1:Add Calculator.** To access the probability functions (i.e., binomial, normal, poisson, geometric, etc.), press (menu), select **5:Probability** and select **5:Distributions**. A menu of probability distributions will be displayed in a list on the left side of the screen.

To calculate a probability for the Normal Distribution, select **2:Normal Cdf** and enter lower bound, upper bound, μ and σ. Press (enter).

To calculate an x-value that corresponds to a given left-sided area under the normal curve, select **3:Inverse Normal.** Fill in the entry boxes (left-sided area, μ and σ). Press (enter).

Sampling Distributions

CHAPTER 8

Section 8.1

Example 5 (pg. 437) Applying the Central Limit Theorem

Weight gain during pregnancy is described as a population with a mean, μ, equal to 30 pounds and a standard deviation, σ, equal to 12.9 pounds. An obstetrician in a low-income neighborhood obtains a sample of 35 patients and calculates a sample average, $\bar{x}$, of 36.2 pounds. What is the probability that the sample average, $\bar{x}$, is 36.2 pounds or higher if, in fact, μ = 30 pounds? Based on the results of this calculation, can the obstetrician conclude that this sample mean of 36.2 pounds is unusual?

Since n > 30, you can conclude that the sampling distribution of the sample mean is approximately normal with $u_{\bar{x}}$ =2716 and $\sigma_{\bar{x}} = 72.8/\sqrt{35}$.

To calculate P($\bar{x} \geq 36.2$), press **2nd** **[DISTR]**, select **2:normalcdf(** and type in **36.2** , **1E99** , **30** , **12.9/$\sqrt{35}$**) and press **ENTER**.

```
normalcdf(36.2,1
E99,30,12.9/√(35
))
      .0022319041
```

▸ Problem 24 (pg. 439) Old Faithful

The times between eruptions of the geyser Old Faithful are normally distributed with μ = 85 minutes and σ = 21.25 minutes.

a.) To find P(X > 95) press **2nd** **[DISTR]**, select **2:normalcdf(** and type in **95** [,] **1 EE 99** [,] **85** [,] **21.25** [)] and press **ENTER**.

b.) To calculate P($\bar{x}$ > 95), press **2nd** **[DISTR]**, select **2:normalcdf(** and type in **95** [,] **1E99** [,] **85** [,] **21.25/**$\sqrt{20}$ [)] and press **ENTER**.

c.) To calculate P($\bar{x}$ > 95), press **2nd** **[DISTR]**, select **2:normalcdf(** and type in **95** [,] **1E99** [,] **85** [,] **21.25/**$\sqrt{30}$ [)] and press **ENTER**.

```
normalcdf(95,1E9
9,85,21.25)
       .3189674148
normalcdf(95,1E9
9,85,21.25/√(20)

       .0176658586
```

```
normalcdf(95,1E9
9,85,21.25/√(30)

       .0049756658
```

◀

Section 8.2

Example 4 (pg. 447) Probabilities of a Sample Proportion

(a.) According to the National Center for Health Statistics, 15% of all Americans have hearing trouble. To calculate the probability that at most 12% of Americans in a sample of 120 have hearing trouble, we need to consider the distribution of the sample proportion, $\hat{p}$. This distribution is approximately normal (provided n*p*(1 - p) $\geq$ 10) with a mean, $\mu_{\hat{p}} = p$, and with standard deviation, $\sigma_{\hat{p}} = \sqrt{\frac{p(1-p)}{n}}$.

To calculate P($\hat{p} \leq .12$), press **2nd [DISTR]**, select **2:normalcdf(** and type in **-1E99** [,] **.12** [,] **.15** [,] $\sqrt{\frac{.15(1-.85)}{120}}$ [)] and press **ENTER**.

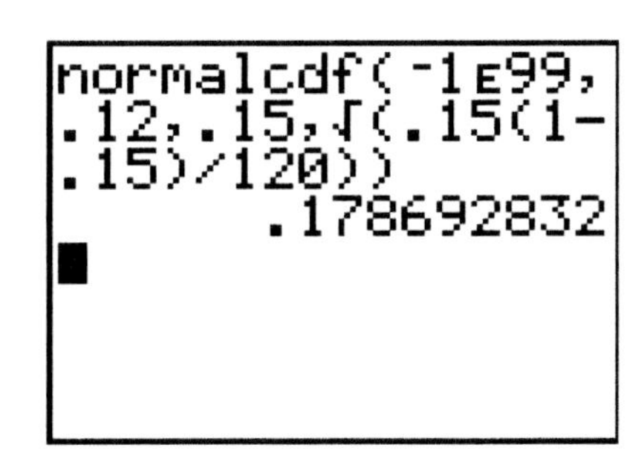

(b.) To determine if it would be unusual to find 26 Americans with hearing trouble in a sample of 120 Americans, we first calculate $\hat{p}$, the sample proportion. In this case, $\hat{p} = \frac{26}{120} = 0.217$. To calculate P($\hat{p} \geq .217$), press **2nd [DISTR]**, select **2:normalcdf(** and type in **.217** [,] **1E99** [,] **.15** [,] $\sqrt{\frac{.15(1-.85)}{120}}$ [)] and press **ENTER**.

```
normalcdf(.217,1
E99,.15,√(.15(1-
.15)/120))
         .0199169327
```

P($\hat{p} \geq .217$) = .0199. Since this probability is very small, this indicates that it would be unusual to find 26 Americans with hearing trouble in a sample of 120.

◀

TI-89 Instructions:

These instructions are designed to give you an overview of sampling distribution calculations using the TI-89.

To calculate a probability involving a sample mean, $\bar{x}$, or a sample proportion, $\hat{p}$, you will use the **Normal Cdf**. In the Stats/List Editor, press **F5** and select **4:Normal Cdf.** Fill in the entry boxes in the **Normal Cdf** screen (lower bound, upper bound, μ and σ.) Note: For the sampling distribution of $\bar{x}$, the standard deviation value that you insert in the data entry box for σ is $\sigma_{\bar{x}}$ which is equal to $\sigma/\sqrt{n}$. For the sampling distribution of $\hat{p}$, the standard deviation value that you insert in the data entry box for σ is $\sigma_{\hat{p}} = \sqrt{\frac{p(1-p)}{n}}$.

TI-*n*spire Instructions:

These instructions are designed to give you an overview of sampling distribution calculations on the TI-*n*spire handhelds.

Press (home) and select **6:New Document**. (Note: If you currently have a document open, the next screen will ask if you want to save the document. Press (tab) to select **No**. Press (enter).) Select **1:Add Calculator.** To calculate a probability involving a sample mean, $\bar{x}$, or a sample proportion, $\hat{p}$, you will use the **Normal Cdf**. Press (menu), select **5:Probability,** select **5:Distributions** and select **2:Normal Cdf**. Fill in the entry boxes in the **Normal Cdf** screen (lower bound, upper bound, μ and σ.) Note: For the sampling distribution of $\bar{x}$, the standard deviation value that you insert in the data entry box for σ is $\sigma_{\bar{x}}$ which is equal to $\sigma/\sqrt{n}$. For the sampling distribution of $\hat{p}$, the standard deviation value that you insert in the data entry box for σ is $\sigma_{\hat{p}} = \sqrt{\frac{p(1-p)}{n}}$.

Estimating the Value Of a Parameter Using Confidence Intervals

Section 9.1

‣ Example 4 (pg. 464) A Confidence Interval for μ (σ known)

Enter the data from Table 1 on pg. 458 into **L1**. Since the sample size is less than 30, we will check for normality using a normal probability plot and we will check for outliers using a Boxplot.

To set up the normal probability plot, press **2nd [STAT PLOT]**. Press ENTER to select **Plot 1**. Highlight **On** and press ENTER. Set **Type** to the normal probability plot which is the third selection in the second row. Press ENTER. Set **Data List** to **L1** and **Data Axis** to **X**. Next, there are three different types of **Marks** that you can select for the graph. The first choice, a small square, is the best one to use.

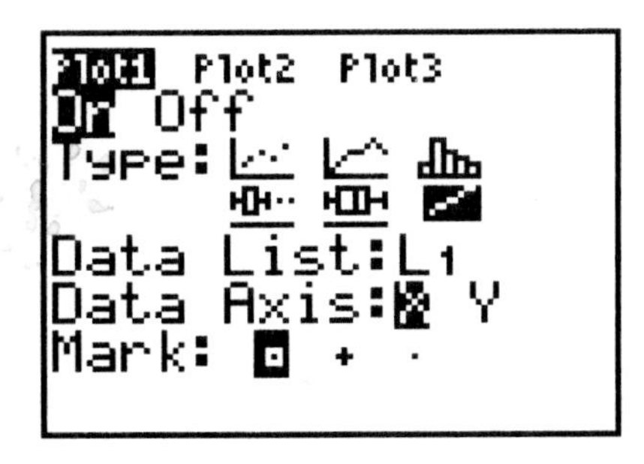

Press ZOOM and select **9:ZoomStat** and ENTER.

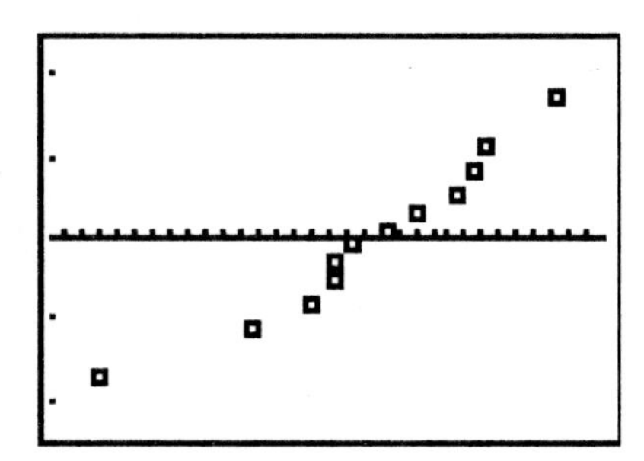

(Note: The calculator draws the normal probability plot along with a horizontal line at the X-axis.) This plot is *fairly* linear, indicating that the data generally follow a normal distribution.

To set up the boxplot, press **2nd [STAT PLOT]**. Press ENTER to select **Plot 1**. Highlight **On** and press ENTER. Set **Type** to the boxplot with outliers which is the first selection in the second row. Press ENTER. Set **XList** to **L1** and **Freq** to **1**. Next, there are three different types of **Marks** that you can select for the graph. The first choice, a small square, is the best one to use.

Press ZOOM and select **9:ZoomStat** and ENTER.

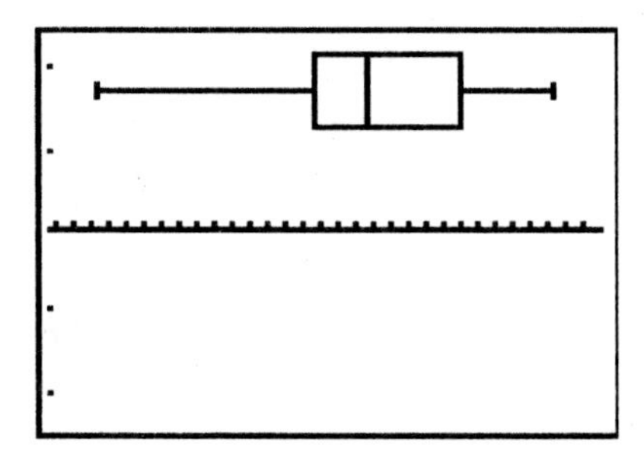

There are no outliers indicated in the boxplot. (Note: Outliers would appear as *'s at the extreme left or right ends of the boxplot.)

Since the data appear to be normally distributed with no outliers, and the population standard deviation is given, the criteria for a Z-interval have been met.

To estimate μ, the population mean, using a 90% confidence interval, press STAT, highlight **TESTS** and select **7:Zinterval.**

```
EDIT CALC TESTS
1:Z-Test...
2:T-Test...
3:2-SampZTest...
4:2-SampTTest...
5:1-PropZTest...
6:2-PropZTest...
7:ZInterval...
```

On the first line of the display, you can select **Data** or **Stats.** For this example, select **Data** because you want to use the actual data which is in **L1**. Press ENTER. Move to the next line and enter 8, the assumed value of σ. On the next line, enter **L1** for **LIST**. For **Freq**, enter **1**. For **C-Level** , enter **.90** for a 90% confidence interval. Move the cursor to **Calculate.**

```
ZInterval
 Inpt:Data Stats
 σ:8
 List:L1
 Freq:1
 C-Level:.90
 Calculate
```

Press ENTER .

```
ZInterval
 (55.818,63.415)
 x̄=59.61666667
 Sx=6.956335678
 n=12
```

A 90% confidence interval estimate of μ , the population mean, is (55.818, 63.415). The output display includes the sample mean (59.617), the sample standard deviation (6.956), and the sample size (12).

▸ Problem 23 (pg. 469)

(a.) A random sample of size n = 25 is selected from a population that is normally distributed with a standard deviation, σ, equal to 13. The sample mean, $\bar{x}$, is equal to 108. To estimate μ, the population mean, using a 96% confidence interval, press STAT, highlight **TESTS** and select **7:Zinterval.**

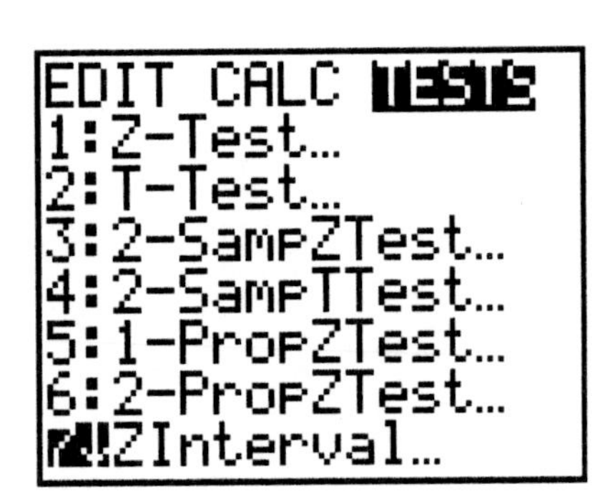

On the first line of the display, you can select **Data** or **Stats.** For this example, select **Stats** because you have the sample mean but not the actual data. Press ENTER. Move to the next line and enter **13**, the value of σ. On the next line, enter **108**, the value for $\bar{x}$, the sample mean. On the next line, enter the sample size, **25**. For **C-Level** , enter **.96** for a 96% confidence interval. Move the cursor to **Calculate.**

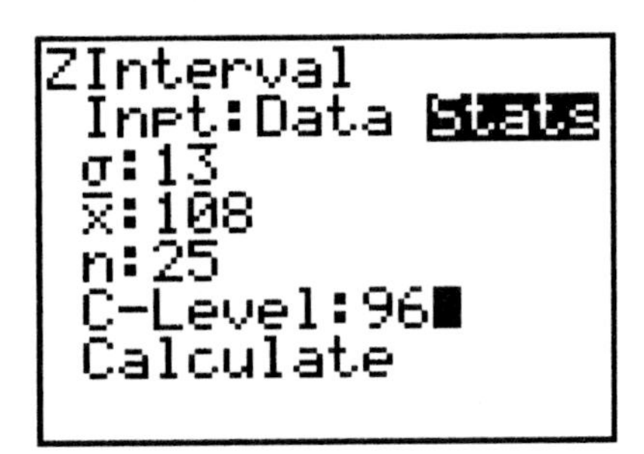

Press ENTER.

```
ZInterval
 (102.66,113.34)
 x̄=108
 n=25
```

(b.) Compute a new confidence interval using n = 10.

(c.) Compute a new confidence interval using n = 25 and a confidence level of 88%.

Section 9.2

Example 4 (pg. 481) A Confidence Interval for μ (σ Unknown)

Enter the data from Table 5 on pg. 479 into **L1**. Since the sample size is less than 30, the first step is to check for normality using a normal probability plot and then to check for outliers using a Boxplot.

To set up the normal probability plot, press **2nd [STAT PLOT]**. Press **ENTER** to select **Plot 1**. Highlight **On** and press **ENTER**. Set **Type** to the normal probability plot which is the third selection in the second row. Press **ENTER**. Set **Data List** to **L1** and **Data Axis** to **X**. Next, there are three different types of **Marks** that you can select for the graph. The first choice, a small square, is the best one to use.

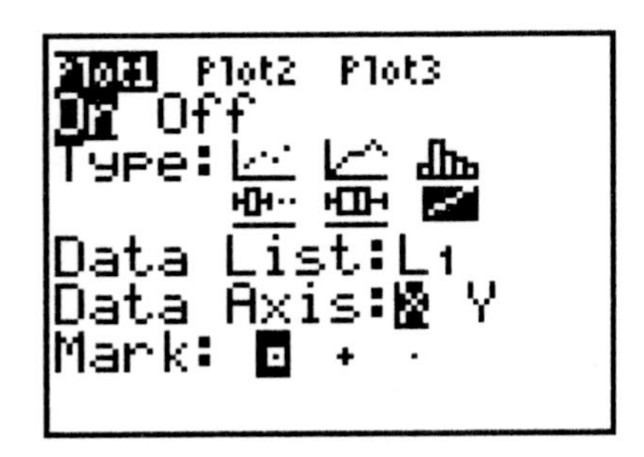

Press **ZOOM** and select **9:ZoomStat** and **ENTER**.

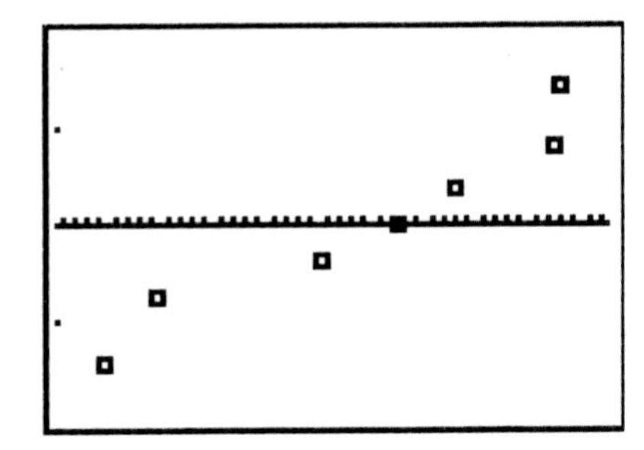

The calculator draws the plot with a horizontal line at the X-axis. This plot is *fairly* linear, indicating that the data generally follow a normal distribution.

To set up the boxplot, press **2nd [STAT PLOT]**. Press **ENTER** to select **Plot 1**. Highlight **On** and press **ENTER**. Set **Type** to the boxplot with outliers which is the first selection in the second row. Press **ENTER**. Set **XList** to **L1** and **Freq** to **1**. Next, there are three different types of **Marks** that you can select for the graph. The first choice, a small square, is the best one to use.

Press **ZOOM** and select **9:ZoomStat** and **ENTER**.

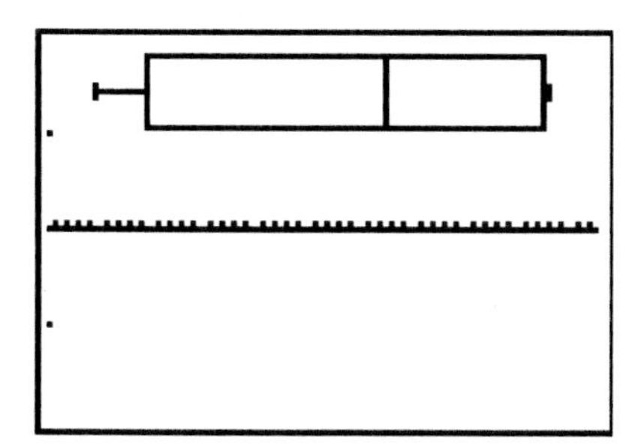

There are no outliers indicated in the boxplot. (Note: Outliers would appear as *'s at the extreme left or right ends of the boxplot.)

In this example, notice that σ is unknown. To construct the confidence interval for μ, the correct procedure under these circumstances ($n < 30$, σ unknown, the population approximately normally distributed and no outliers) is to use a T-Interval.

Press **STAT**, highlight **TESTS**, scroll through the options and select **8:TInterval** and press **ENTER** . Select **Data** for **Inpt** and press **ENTER**. For **List**, enter **L1** and for **Freq**, enter **1**. Set **C-level** to **.95**. Highlight **Calculate**.

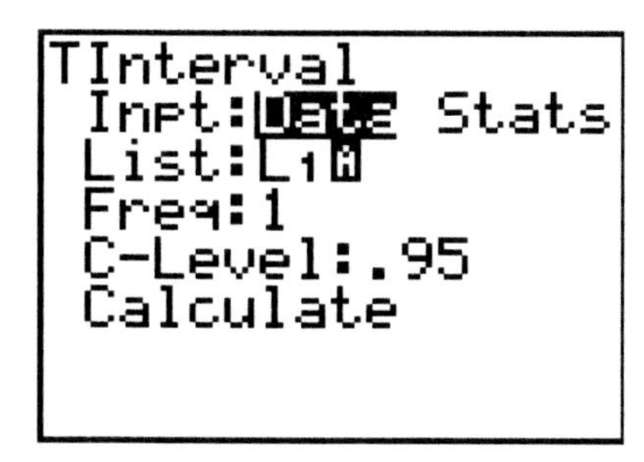

Press **ENTER**.

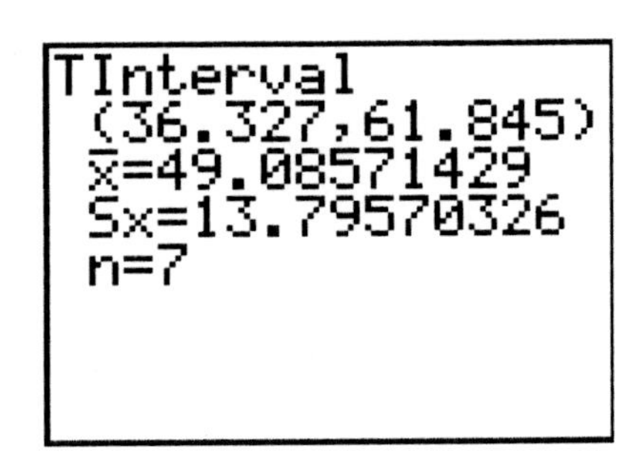

A 95% confidence interval for μ is (36.327, 61.845). The sample statistics (mean, standard deviation and sample size) are also given in the output screen.

◀

Problem 11 (pg. 483)

(a.) In this example, $\bar{x}$ = 18.4, s= 4.5 and the sample size, n, = 35.
Since σ, the population standard deviation is unknown, the correct procedure for constructing a confidence interval for μ is the T-procedure.

Press **STAT**, highlight **TESTS**, scroll through the options and select **8:TInterval** and press **ENTER**. In this example, you do not have the actual data. What you have are the summary statistics of the data, so select **Stats** and press **ENTER**. Enter the values for $\bar{x}$, **Sx** and **n.** Enter **.95** for **C-level.** Highlight **Calculate**.

```
TInterval
 Inpt:Data Stats
 x̄:18.4
 Sx:4.5
 n:35
 C-Level:95
 Calculate
```

Press **ENTER**.

```
TInterval
 (16.854,19.946)
 x̄=18.4
 Sx=4.5
 n=35
```

A 95% confidence interval estimate for μ is (16.854, 19.946).

(b.) Compute a new confidence interval using n = 50.

(c.) Compute a new confidence interval using n = 35 and a confidence level of 99%.

▸ Problem 25 (pg.485)

Enter the data into **L1**. Since the sample size is less than 30, the first step is to check for normality using a normal probability plot and then to check for outliers using a Boxplot.

To set up the normal probability plot, press **2nd [STAT PLOT]**. Press ENTER to select **Plot 1**. Highlight **On** and press ENTER. Set **Type** to the normal probability plot which is the third selection in the second row. Press ENTER. Set **Data List** to **L1** and **Data Axis** to **X**. For **Marks** select the small square.

Press ZOOM and select **9:ZoomStat** and ENTER.

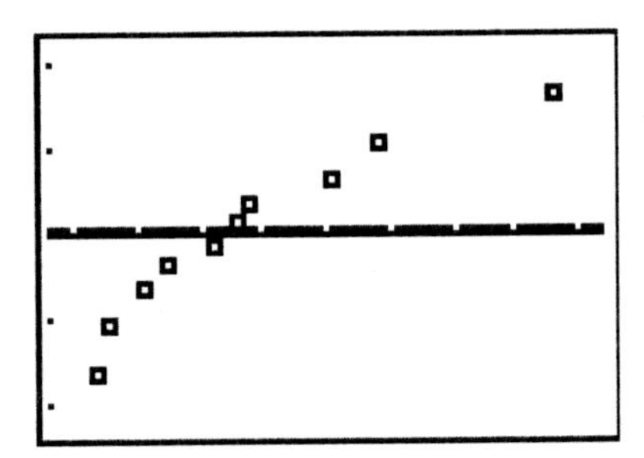

This plot is *fairly* linear, indicating that the data generally follow a normal distribution.

To set up the boxplot, press **2nd** [STAT PLOT] . Press ENTER to select **Plot 1**. Highlight **On** and press ENTER. Set **Type** to the boxplot with outliers which is the first selection in the second row. Press ENTER. Set **XList** to **L1** and **Freq** to **1**. For **Marks** select the small square.

Press ZOOM and select **9:ZoomStat** and ENTER.

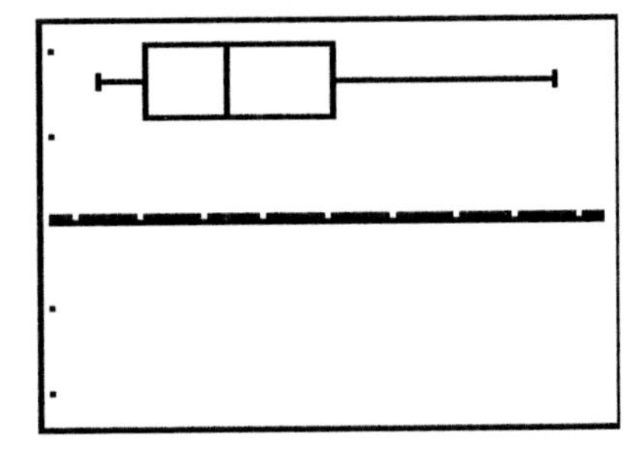

There are no outliers indicated in the boxplot. (Note: Outliers would appear as *'s at the extreme left or right ends of the boxplot.)

Since the data appear to be normally distributed with no outliers, and the population standard deviation is unknown, the criteria for the T-interval have been met.

Press STAT, highlight **TESTS** and select **8:T-interval**. Select **Data** for **Inpt** and press ENTER. For **List**, enter **L1** and for **Freq**, enter **1**. Set **C-level** to **.95**. Highlight **Calculate.**

```
TInterval
 Inpt:Data Stats
 List:L1
 Freq:1
 C-Level:95
 Calculate
```

Press ENTER.

```
TInterval
 (151.85,183.15)
 x̄=167.5
 Sx=21.87972171
 n=10
```

Section 9.3

▸ Example 3 (pg. 490) A Confidence Interval for a Population Proportion

In this example, 935 teenagers were asked whether they participate in any form of content creation for the Internet. 598 individuals responded 'Yes.' Construct a 95 % confidence interval for p, the true proportion of all teenagers who participate in content creation for the Internet.

Press **STAT**, highlight **TESTS**, scroll through the options and select **A:1-PropZInt**. The value for X is the number of teenagers in the sample of 935 who responded 'Yes', so **X = 598**. The number who were surveyed is n, so **n = 935**. Enter **.95** for **C-level**.

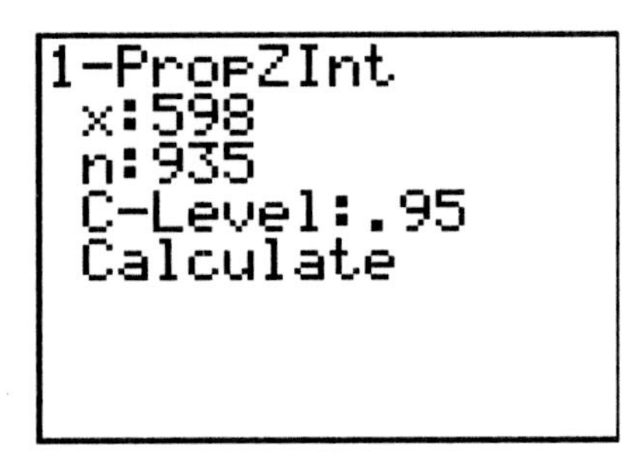

Highlight **Calculate** and press **ENTER**.

```
1-PropZInt
 (.6088,.67035)
 p̂=.6395721925
 n=935
```

In the output display the confidence interval for p is (.6088, .67035). The sample proportion, $\hat{p}$, is .63957 and the number surveyed is 935.

Note: You should calculate $n * \hat{p} * (1 - \hat{p})$. This value must be greater than or equal to 10 in order to use this confidence interval procedure. (It is actually easier to do this calculation after you have calculated the confidence interval because the calculator displays the value of $\hat{p}$ as part of the output.). For this example, the calculation is 935*.64*.36. This value is greater than 10, so this supports the use of the confidence interval procedure. Also, make sure that the sample size is no more than 5% of the population size. In this example, because the population is so large (all teenagers), the sample size criteria is certainly satisfied.

◀

TI-89 Instructions:

These instructions are designed to give you an overview of normal probability plots and one-sample confidence intervals on the TI-89.

Normal Probability Plot
To construct a normal probability plot, first check the y-registers by pressing ◆ and **Y=** and clearing all the y-functions. Return to the **Stats/List Editor**, clear all the lists and enter the data into **list1**. Press **F2** and select **2:Norm Prob Plot**. On the next screen, scroll to the List entry and use **2nd** **VAR-LINK** to select **list1.** Press **ENTER**. A column labeled **zscores** has been created in the **List Editor.** Press **F2** and select **1:Plot Setup**. On the Plot Setup screen press **F5** to display the normal probability plot.

Confidence Interval
To create a confidence interval, in the Stats/List Editor press **2nd** **F7** and select the appropriate interval. For **1:ZInterval** or **2:TInterval,** the next step is to choose the Data Input Method. Choose **Stats** if you are using the summary statistics (sample mean and standard deviation). Choose **Data** if you have entered the actual data values into a list. On the next screen, fill in all the information and click **ENTER** until the confidence interval and summary statistics are displayed.

TI-*n*spire Instructions:

These instructions are designed to give you an overview of normal probability plots and one-sample confidence intervals on the TI-*n*Spire handheld.

Normal Probability Plot
Press (home) and select **6:New Document**. (Note: If you currently have a document open, the next screen will ask if you want to save the document. Press (tab) to select **No**. Press (enter).) Select **3:Add Lists & Spreadsheet. M**ove to the top of column **A** and in the box next to '**A**' type in a name. Move to **Line 1** in column **A** and begin entering your data values.

To construct a normal probability plot, press (home) and select **5: Data & Statistics**. A new page will open with a scatter of points. Press and hold the down arrow to move the cursor to the middle of the x-axis at the bottom of the screen. When the box '**Click to add variable**' appears, press (enter). Highlight the Column **A** name and press (enter). A **Dotplot** of the data in Column A will be displayed. Press (menu), select **1:Plot Type** and select **4: Normal Probability Plot.** A normal probability plot will be displayed. Press (ctrl) and the left arrow to return to the Spreadsheet.

Confidence Interval
To create a confidence interval, on the Spreadsheet page, press (menu), select **4:Statistics, 3:Confidence Intervals** and select the appropriate interval. For **1:ZInterval** or **2:tInterva**l, use the down-arrow to choose the **Data Entry Method**. Choose **Stats** if you are using the summary statistics (sample mean and standard deviation). Choose **Data** if you have entered the actual data values into a list. Press (tab) to highlight **OK** and press (enter). Fill in the entry boxes on the next screen. Press (tab) to move through the entry boxes. The output will be displayed in two columns in the spreadsheet. To expand a column width, move to any location in that particular column and press (menu), select **1:Actions, 2:Resize** and **1:Resize Column Width.** The column is now highlighted. By clicking the right arrow you can expand the column. Press (enter). Press the down arrow to remove the highlighting.

Hypothesis Tests Regarding a Parameter

Section 10.2

▸ Example 3 (pg. 530) Testing a Claim about μ (σ known): Large Sample

In this example, we are given the sample statistics: $\bar{x} = 540$ and n = 40. Since n > 30, the Central Limit Theorem applies so there is no need to test for normality or for the presence of outliers.

The hypothesis test, $H_o: \mu = 515$ vs. $H_a: \mu > 515$, is a right-tailed test. Since, the population standard deviation, σ, is given, the **Z-Test** is the appropriate test. To run the test, press **STAT**, highlight **TESTS** and select **1:Z-Test**. Since we are using the sample statistics for the analysis, select **Stats** for **Inpt** and press **ENTER**. For μ_0 enter 515, the value for μ in the null hypothesis. For σ enter 114, for $\bar{x}$ enter 540 and for n enter 40. On the next line, choose the appropriate alternative hypothesis and press **ENTER**. For this example, it is > μ_0, a right-tailed test.

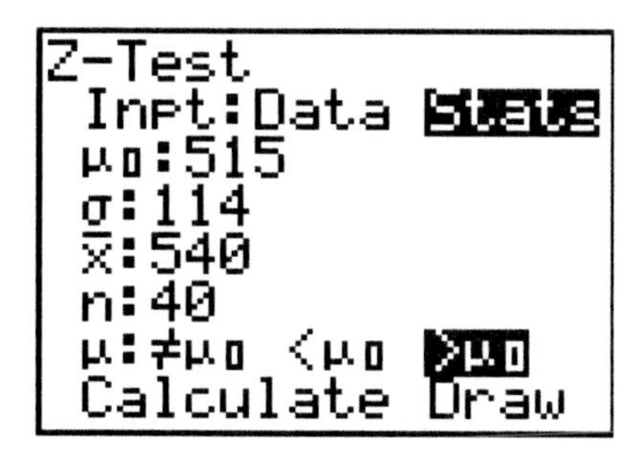

There are two choices for the output of this test. The first choice is **Calculate**. The output displays the alternative hypothesis, the calculated z-value, the P-value, $\bar{x}$ and n.

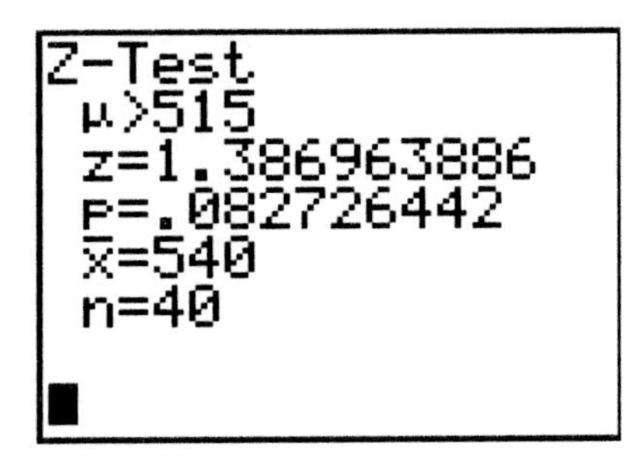

Since p = .0827, which is greater than α, the correct conclusion is to **Fail to Reject** H_o. (Note: The P-value calculated using the TI-84 is slightly different from the P-value obtained using the Z-table. That difference is simply due to rounding.)

To view the second output option, clear all the previous drawings by pressing 2nd **[DRAW]** and selecting **1:ClrDraw** and pressing ENTER ENTER. Press 2nd **[STATPLOT]** and TURN OFF all PLOTS. Make sure that there is nothing stored in the Y-registers. Press Y= and check the Y-registers. If any of them contain a function, move the cursor to that Y-register and press CLEAR. Now press STAT, highlight **TESTS**, and select **1:Z-Test**. All the necessary information for this example is still stored in the calculator. Scroll down to the bottom line and select **DRAW**. A normal curve is displayed with the right-tail area of .0827 shaded. This shaded area is the area to the right of the calculated Z-value. The Z-value and the P-value are also displayed.

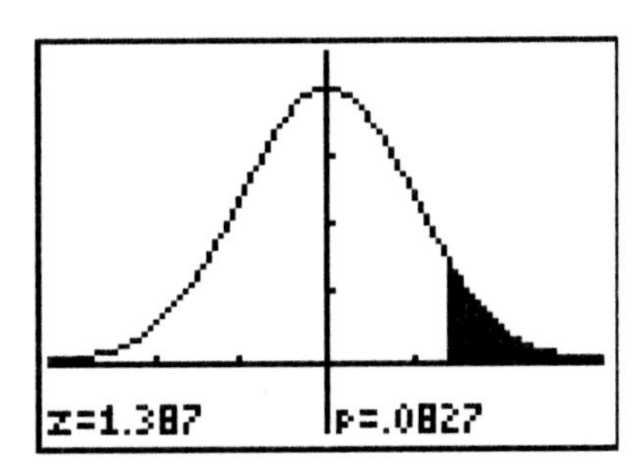

▸ Example 5 (pg. 533) Testing a Claim about μ (σ known): Small Sample

Enter the data from Table 2 on pg. 532 into L1. Because the sample size is less than 30, the data must be tested for normality and checked for outliers.

To set up the normal probability plot, press **2nd** **[STAT PLOT]** . Press **ENTER** to select **Plot 1**. Highlight **On** and press **ENTER**. Set **Type** to the normal probability plot which is the third selection in the second row. Press **ENTER**. Set **Data List** to **L1** and **Data Axis** to **X**. For **Marks** select the small square.

Press **ZOOM** and select **9:ZoomStat** and **ENTER**.

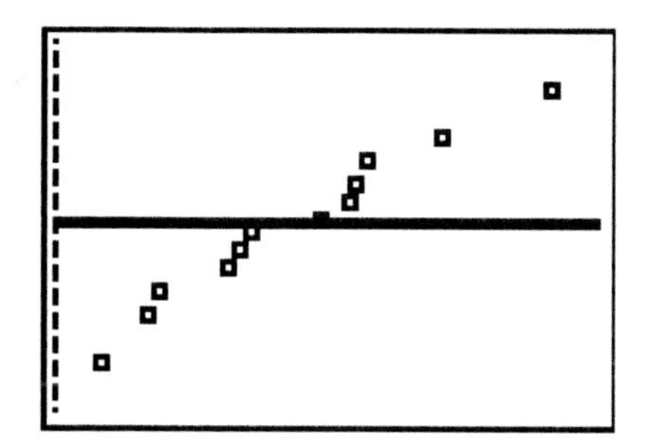

This plot is *fairly* linear, indicating that the data generally follow a normal distribution.

To set up the boxplot, press **2nd** **[STAT PLOT]**. Press **ENTER** to select **Plot 1**. Highlight **On** and press **ENTER**. Set **Type** to the boxplot with outliers which is the first selection in the second row. Press **ENTER**. Set **XList** to **L1** and **Freq** to **1**. For **Marks** select the small square.

Press **ZOOM** and select **9:ZoomStat** and **ENTER**.

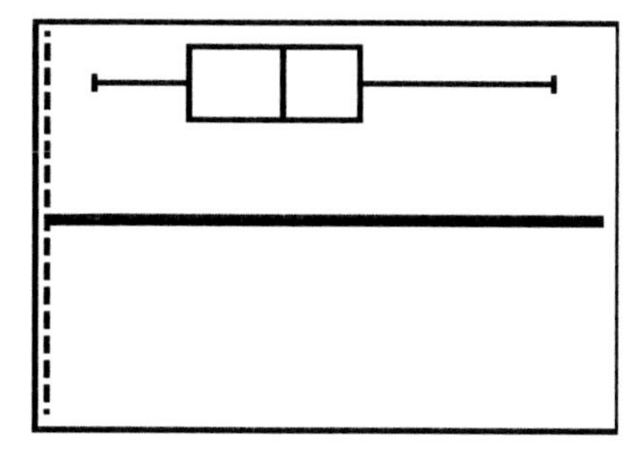

There are no outliers indicated in the boxplot. (Note: Outliers would appear as *'s at the extreme left or right ends of the boxplot.)

The hypothesis test, $H_o: \mu = 50.64$ vs. $H_a: \mu \neq 50.64$, is a two-tailed test. Since the population standard deviation, σ, is given, the **Z-Test** is the appropriate test. To run the test, press **STAT**, highlight **TESTS** and select **1:Z-Test**. Since you are using the actual data, which is stored in L1, for the analysis, select **Data** for **Inpt** and press **ENTER**. For μ_0 enter 50.64, the value for μ in the null hypothesis. For σ enter 18.49. Enter **L1** for **List,** and **1** for **Freq**. On the next line, choose the appropriate alternative hypothesis and press **ENTER**. For this example, it is $\neq \mu_0$, a two-tailed test.

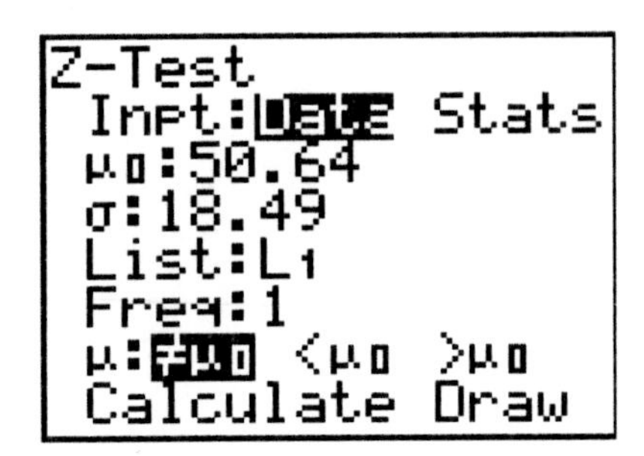

There are two choices for the output of this test. The first choice is **Calculate**. The output displays the alternative hypothesis, the calculated z-value, the P-value, $\bar{x}$ and n.

```
Z-Test
 μ≠50.64
 z=2.693000215
 p=.007081319
 x̄=65.01416667
 Sx=25.45873362
 n=12
```

Since p = .007, which is less than α, the correct conclusion is to **Reject** H_o.

To view the second output option, clear all the previous drawings by pressing 2nd **[DRAW]** and selecting **1:ClrDraw** and pressing **ENTER** **ENTER**. Press 2nd **[STATPLOT]** and TURN OFF all PLOTS. Make sure that there is nothing stored in the Y-registers. Press **Y=** and check the Y-registers. If any of them contain a function, move the cursor to that Y-register and press **CLEAR**. Now press **STAT**, highlight **TESTS**, and select **1:Z-Test**. All the necessary information for this example is still stored in the calculator. Scroll down to the bottom line and select **DRAW**. A normal curve is displayed with each tail area of .0035 shaded. (Note: Because the areas are so small in this example, they are not really visible in the curve.) The shaded areas total to 0.007 which is the P-value. The Z-value and the P-value are also displayed.

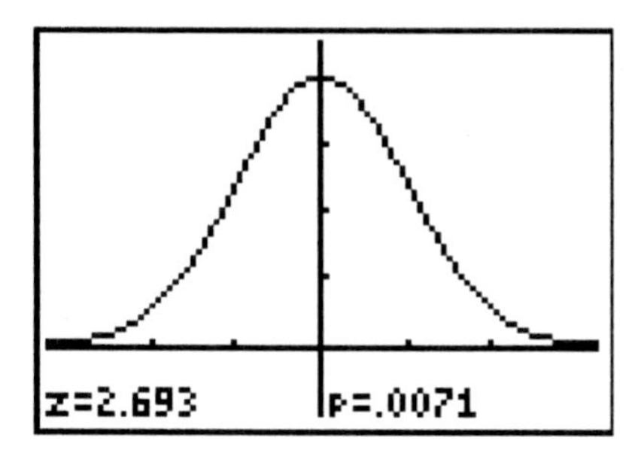

◀

▸ Example 6 (pg. 534) Using a Confidence Interval to Test a Claim

Enter the data from Table 2 on pg. 532 into L1. Because the sample size is less than 30, the data must be tested for normality and checked for outliers. (Note: These tests were done with the previous Example and the results indicated that the data were normally distributed with no outliers.)

To estimate μ, the population mean, using a 95% confidence interval, press STAT, highlight **TESTS** and select **7:Zinterval.**

```
EDIT CALC TESTS
1:Z-Test...
2:T-Test...
3:2-SampZTest...
4:2-SampTTest...
5:1-PropZTest...
6:2-PropZTest...
7:ZInterval...
```

On the first line of the display select **Data.** Press ENTER. Move to the next line and enter 18.49, the assumed value of σ. On the next line, enter **L1** for **LIST**. For **Freq**, enter **1**. For **C-Level**, enter **.95** for a 95% confidence interval. Move the cursor to **Calculate.**

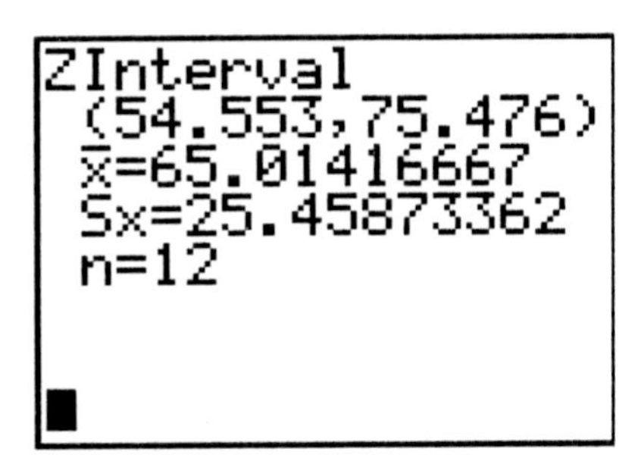

The 95% confidence interval for μ is (54.553, 75.476). Notice that this confidence interval does not contain the hypothesized value for μ (50.64). Since the hypothesized value is not contained in the confidence interval, the correct decision is: **Reject the null hypothesis.** There is sufficient evidence at the .05 significance level to support the claim that the mean monthly cell phone bill is different from $50.64. ◂

Problem 15 (pg. 536)

Test the hypotheses: $H_o: \mu = 20$ vs. $H_a: \mu < 20$. The underlying population is assumed to be normally distributed with σ = 3. The sample mean, $\bar{x}$, = 18.3, and n = 18. Press **STAT**, highlight **TESTS** and select **1:Z-Test**. For **Inpt**, choose **Stats** and press **ENTER**. Fill the input screen with the appropriate information. Choose < μ_0 for the alternative hypothesis and press **ENTER**.

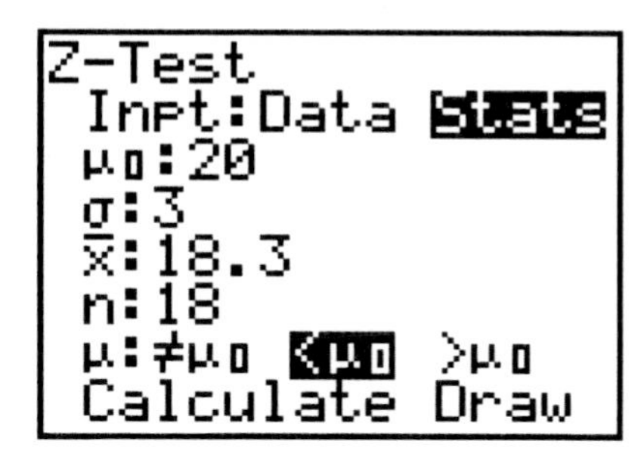

Highlight **Calculate** and press **ENTER**.

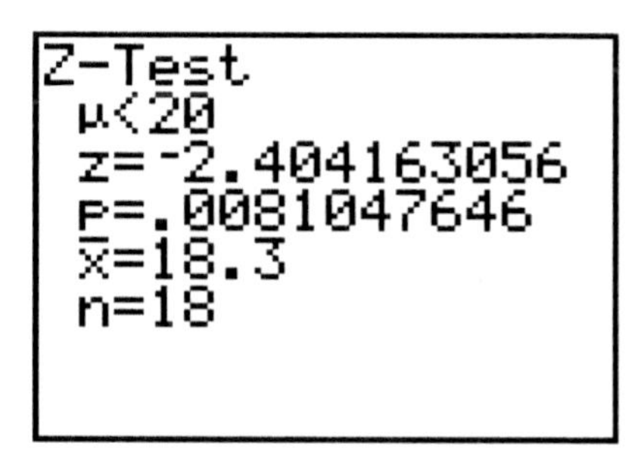

Or, (after clearing all graphs and Y-registers) highlight **Draw** and press **ENTER**.

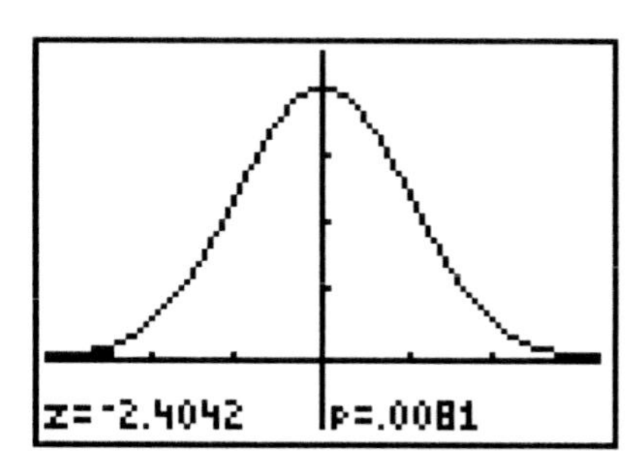

The P-value is .008. Since the P-value is less than α, the correct conclusion is to **Reject** H_o.

▸ Problem 17 (pg. 536)

Test the hypotheses: $H_o: \mu = 105$ vs. $H_a: \mu \neq 105$. In this example, the sample size is greater than 30. Since the sample size is *large,* the Central Limit Theorem applies and we can assume that the sampling distribution of $\bar{x}$ is approximately normal. The population standard deviation, σ, is equal to 12.

(b.) To run the test, press **STAT**, highlight **TESTS** and select **1:Z-Test**. For **Inpt**, choose **Stats** and press **ENTER**. Fill the input screen with the appropriate information. Choose $\neq \mu_0$ for the alternative hypothesis and press **ENTER**.

Highlight **Calculate** and press **ENTER**.

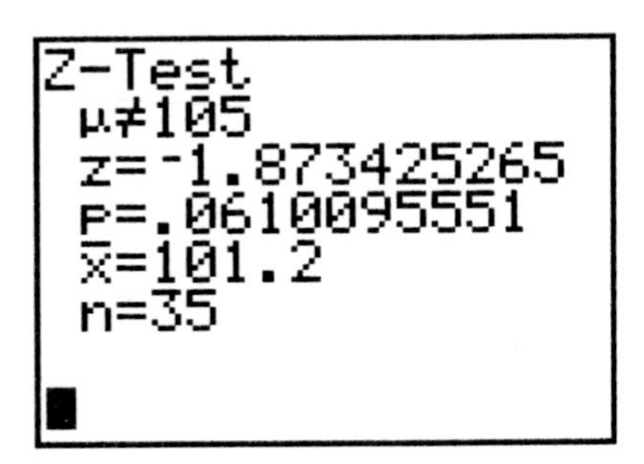

Or, (after clearing all graphs and Y-registers) highlight **Draw** and press **ENTER**.

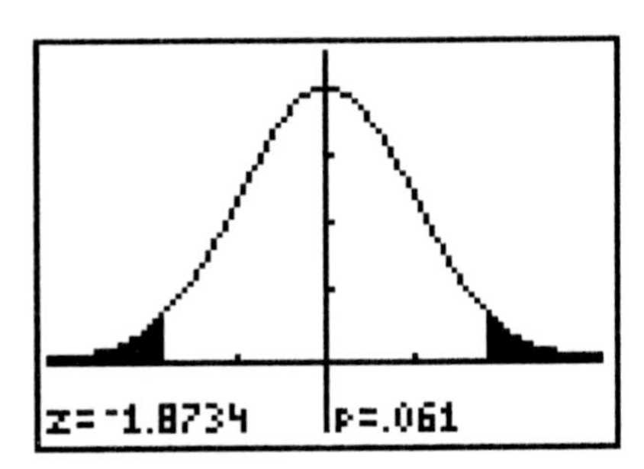

The P=value is .061. Since the P-value is greater than α, the correct conclusion is to **Fail to Reject** H_o.

(d.) Using a **ZInterval**, construct a 95% confidence interval and compare the confidence interval to the results of the hypothesis test.

◂

Section 10.3

Example 2 (pg. 545) Testing a Claim about μ (σ unknown): Large Sample

Refer to Example1 on pg. 543. Test the hypotheses: $H_o : \mu = 18.1$ vs. $H_a : \mu <$ 18.1. In this example, the sample size is greater than 30. Since the sample size is *large,* the Central Limit Theorem applies and we can assume that the sampling distribution of $\bar{x}$ is approximately normal. Also, notice that σ, the population standard deviation is not given. Instead the sample standard deviation is given; therefore, the appropriate test is the T-test.

To run the test, press **STAT**, highlight **TESTS** and select **2:T-Test**. For **Inpt**, choose **Stats** and press **ENTER**. Fill the input screen with the appropriate information. (μ_0=18.1, $\bar{x}$=16.8, s = 4.7 and n = 40.) Choose < μ_0 for the alternative hypothesis and press **ENTER**. Highlight **Calculate** and press **ENTER**.

```
T-Test
 μ<18.1
 t=-1.749345089
 p=.0440489391
 x̄=16.8
 Sx=4.7
 n=40
```

Or, (after clearing all graphs and the Y-registers) highlight **Draw** and press **ENTER**.

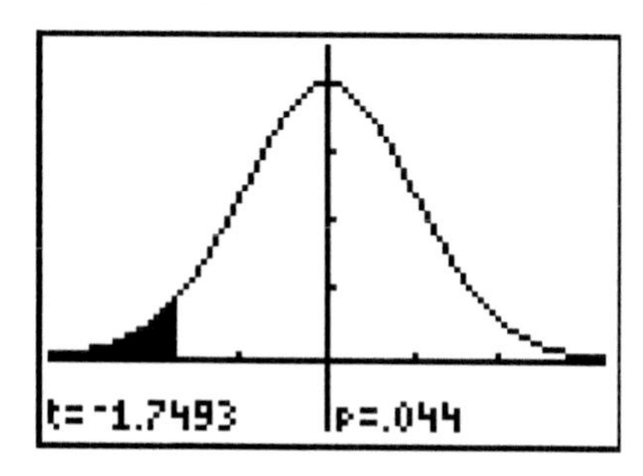

The P=value is .044. Since the P-value is less than α, the correct conclusion is to **Reject** H_o.

▸ Problem 9 (pg. 548)

Test the hypotheses: $H_o: \mu = 20$ vs. $H_a: \mu < 20$. The underlying population is known to be normally distributed. The sample statistics are $\bar{x} = 18.3$, s = 4.3 and n = 18. Press **STAT**, highlight **TESTS** and select **1:T-Test**. For **Inpt**, choose **Stats** and press **ENTER**. Fill the input screen with the appropriate information. Choose < μ_0 for the alternative hypothesis and press **ENTER**.

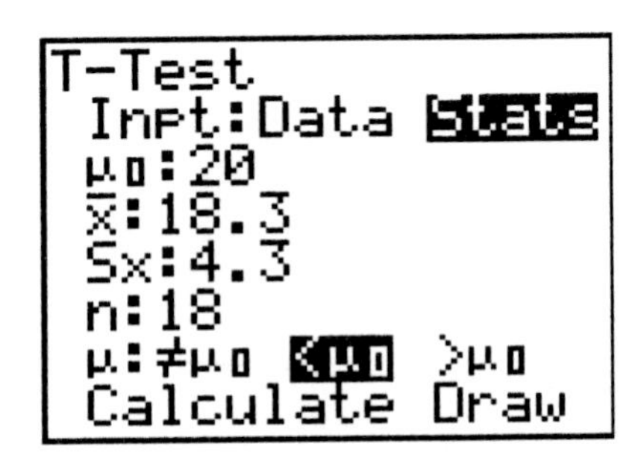

Highlight **Calculate** and press **ENTER**.

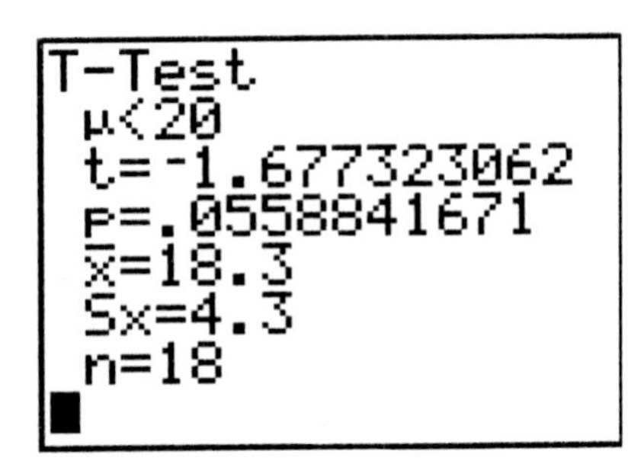

Or, (after clearing all graphs and the Y-registers) highlight **Draw** and press **ENTER**.

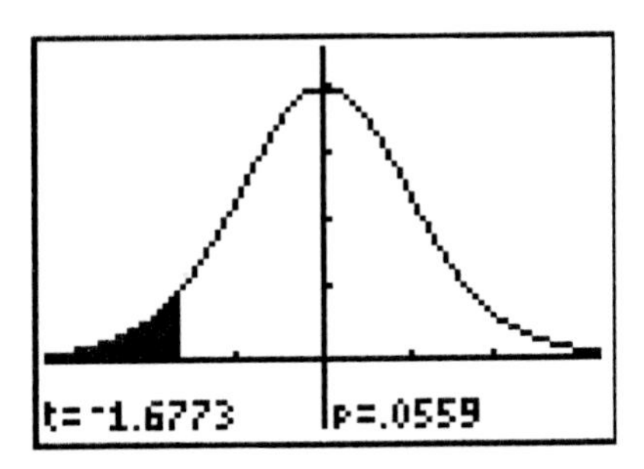

The P=value is .0559. Since the P-value is greater than α, the correct conclusion is to **Fail to Reject** H_o.

▸ Problem 25 (pg. 550)

Enter the data into L1. Because the sample size is less than 30, the data must be tested for normality and checked for outliers.

To set up the normal probability plot, press **2nd** [STAT PLOT] . Press ENTER to select **Plot 1**. Highlight **On** and press ENTER. Set **Type** to the normal probability plot which is the third selection in the second row. Press ENTER. Set **Data List** to **L1** and **Data Axis** to **X**. For **Marks** select the small square.

Press ZOOM and select **9:ZoomStat** and ENTER.

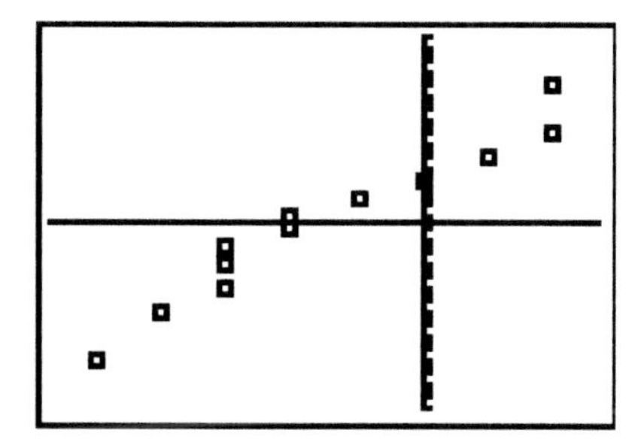

This plot is *fairly* linear, indicating that the data generally follow a normal distribution.

To set up the boxplot, press **2nd [STAT PLOT]**. Press ENTER to select **Plot 1**. Highlight **On** and press ENTER. Set **Type** to the boxplot with outliers which is the first selection in the second row. Press ENTER. Set **XList** to **L1** and **Freq** to **1**. For **Marks** select the small square.

Press ZOOM and select **9:ZoomStat** and ENTER.

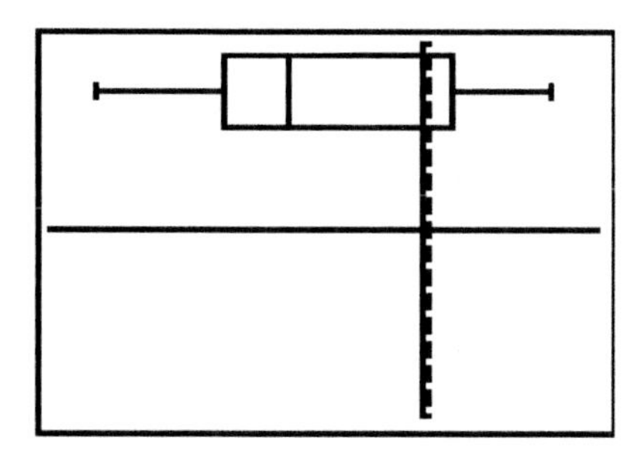

There are no outliers indicated in the boxplot. (Note: Outliers would appear as *'s at the extreme left or right ends of the boxplot.)

This test is a left-tailed test of $H_o: \mu = 0$ vs. $H_a: \mu \neq 0$. Since n < 30, and the population standard deviation, σ, is unknown, the T-Test is the appropriate test. This test requires the underlying population to be approximately normally distributed with no outliers, as was verified in the plots.

Press STAT, highlight **TESTS** and select **2:T-Test**. Choose **Data** for **Inpt** and press ENTER. Fill in the following information: μ_0 **= 0, List = L1,** and **Freq =1**. Choose the right-tailed alternative hypothesis, $\neq \mu_0$, and press ENTER. Highlight **Calculate** and press ENTER

```
T-Test
 μ≠0
 t=-2.243635065
 p=.0464047507
 x̄=-.015
 Sx=.0231595258
 n=12
```

Or, highlight **Draw** and press ENTER.

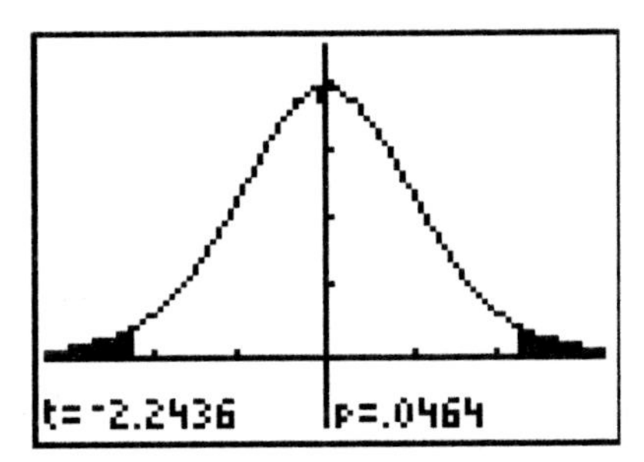

Since the P-value is less than α, the correct conclusion is to **Reject** H_o.

Section 10.4

Example 3 (pg. 556) Testing a Claim about a Population Proportion: Large Sample

Refer to Example 2 on pg. 555. This hypothesis test is a two-tailed test of: $H_o: p = .135$ vs. $H_a: p \neq .135$. The procedure that is used for this test is the **1-Proportion Test**. This test has two requirements. The first requirement is: $n * p_0 * (1 - p_0) \geq 10$. To verify this, calculate 710*.135*(1-.135). The result is greater than 10, so the first requirement is satisfied. The second requirement is that the sample size is not more than 5% of the *population* size. In this example, the population is *all babies between 12 and 15 months of age.* We don't know the exact size of the population, but it is in the millions. The sample size of 710 is definitely less than 5% of the population size.

To run the test, press **STAT**, highlight **TESTS** and select **5:1-PropZTest**. This test requires a value for p_0, which is the value for p in the null hypothesis. Enter **.135** for p_0. Next, a value for X is required. X is the number of "successes" in the sample. In this example, a success is " experiencing a loss of appetite", so **X** is equal to **121**. Next, enter the value for **n**. Select $\neq p_0$ for the alternative hypothesis and press **ENTER**.

```
1-PropZTest
 p0:.135
 x:121
 n:710
 prop≠p0 <p0 >p0
 Calculate Draw
```

Highlight **Calculate** and press **ENTER**.

```
1-PropZTest
 prop≠.135
 z=2.762064888
 p=.0057438195
 p̂=.1704225352
 n=710
```

The output displays the alternate hypothesis that was selected, the calculated Z-value, the P-value, the sample proportion, $\hat{p}$, and the sample size. (Note: $\hat{p}$ = 121/710.)

Or, (after clearing all graphs and Y-registers) highlight **Draw** and press **ENTER**.

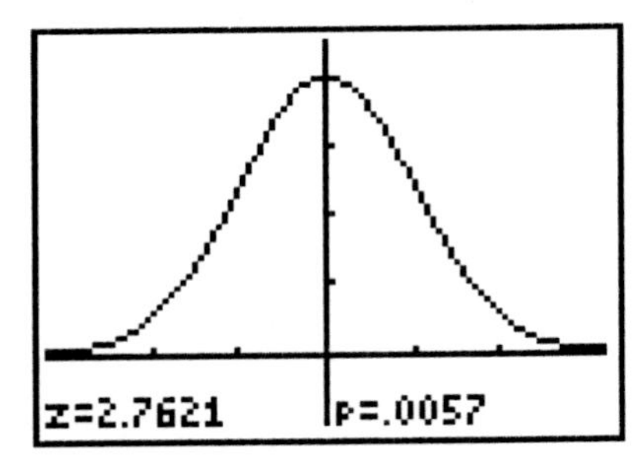

Since the P-value is less than α, the correct conclusion is to **Reject** H_o.

◀

▸ Example 4 (pg. 557) Testing a Claim about a Population Proportion: Small Sample

In this test of a population proportion, the requirement $n * p_0 * (1 - p_0) \geq 10$ is not satisfied. (The calculation 35*.489*(1-.489) is equal to 8.75.) (Note: In cases in which the sample size is relatively small this requirement is often not satisfied.)

An alternative method of testing a hypothesis about a population proportion is to use the binomial probability formula to calculate the likelihood of the sample result. If the sample result is *unusual* then we will **reject the null hypothesis**. We define *unusual* events as events that have a probability less than .05.

The hypothesis test is: $H_o: p = .489$ vs. $H_a: p > .489$. The sample statistics are n = 35 and X = 21. Using the binomial probability formula, we calculate the likelihood of obtaining 21 or more males who consume the recommended daily allowance of calcium in the sample of 35. We assume that the proportion of males in the population who consume the recommended daily allowance of calcium is .489.

Press **1 – 2nd DISTR** and select **B:binomcdf(**. Type in **35 , .489 , 20).**

```
1-binomcdf(35,.4
89,20)
      .1261068497
```

(Note: The command **binomcdf** calculates the probability that X ≤ 20, which is the *complement* of the probability that X ≥ 21. To obtain P(X ≥ 21), we calculate P(X ≤ 20) and subtract this value from 1.)

The result is: P(X ≥ 21) = .126. Since this probability is greater than .05, the correct conclusion is to **Fail to Reject** H_o.

◂

▸ Problem 20 (pg. 559)

(pg. 559)

This hypothesis test is a right-tailed test of: $H_o: p = .85$ vs. $H_a: p > .85$. To use the **1-Proportion Test,** first you must determine whether the requirements for this test have been satisfied. The first requirement is: $n * p_0 * (1 - p_0) \geq 10$. To verify this, calculate 200*.85*(1-.85). The result is greater than 10, so the first requirement is satisfied. The second requirement is that the sample size is not more than 5% of the *population* size. In this example, the population is *all American adults.* We don't know the exact size of the population, but it is in the millions. The sample size of 150 is definitely less than 5% of the population size.

To run the test, press **STAT**, highlight **TESTS** and select **5:1-PropZTest**. Enter **.85** for p_0. For **X,** enter **171** and for n, enter **200**. Select $> p_0$ for the alternative hypothesis and press **ENTER**.

Highlight **Calculate** and press **ENTER**.

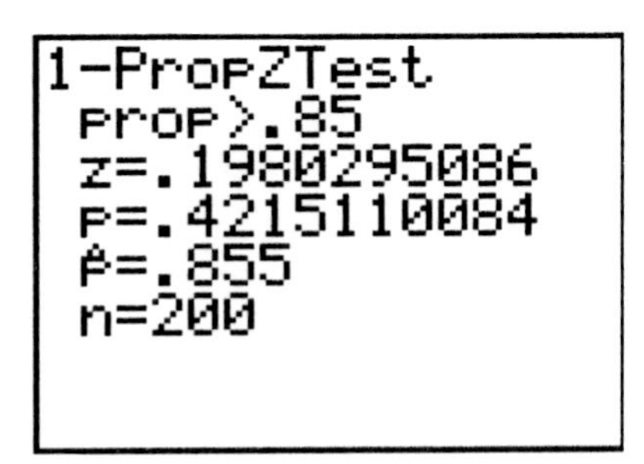

Or highlight **Draw** and press **ENTER**.

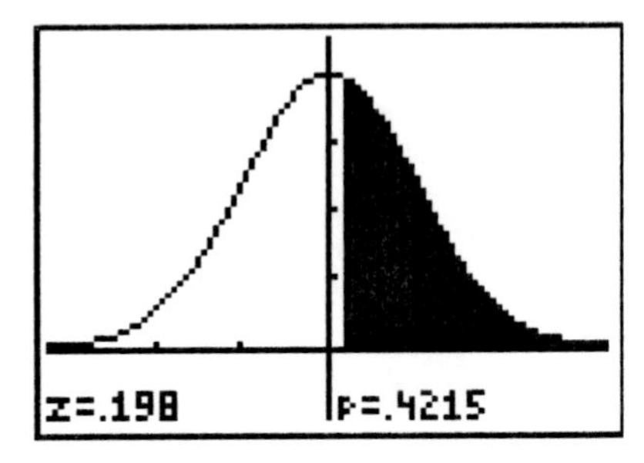

Since the P-value is greater than α, the correct conclusion is to **Fail to Reject** H_o. The data does not support the nutritionist's claim that the percentage of American adults who eat salad at least once a week is higher than 85%. ◂

▸ Problem 21 (pg. 559)

In this test of a population proportion, the requirement $n * p_0 * (1 - p_0) \geq 10$ is not satisfied. (The calculation 120*.04*(1-.04) is equal to 4.608.) (Note: In cases in which the sample size is relatively small this requirement is often not satisfied.)

An alternative method of testing a hypothesis about a population proportion is to use the binomial probability formula to calculate the likelihood of the sample result. If the sample result is *unusual* then we will **reject the null hypothesis**. We define *unusual* events as events that have a probability less than .05.

The hypothesis test is: $H_o: p = .04$ vs. $H_a: p < .04$. The sample statistics are n = 120 and X = 3. Using the binomial probability formula, we calculate the likelihood of obtaining 3 or fewer mothers who smoked 21 or more cigarettes during pregnancy. We assume that the proportion of mothers who smoked 21 or more cigarettes during pregnancy is .04.

Press **2nd DISTR** and select **A:binomcdf(.** Type in **120 , .04 , 3).**

```
binomcdf(120,.04
,3)
          .288658855
```

The result is: P(X ≤ 3) = .2887. Since this probability is greater than .05, the correct conclusion is to **Fail to Reject** H_o. The data does not support the obstetrician's belief that less than 4% of mothers smoked 21 or more cigarettes during pregnancy.

◂

Section 10.5

Example 2 (pg. 564) Testing a Claim About a Population Standard Deviation

This is a left-tailed hypothesis test about σ. The hypotheses are: H_o: $\sigma = 0.75$ vs. $H_a: \sigma < 0.75$.

Enter the data from Table 4 on pg. 563 into L1. Because the sample size is less than 30, the data must be tested for normality.

To set up the normal probability plot, press **2nd [STAT PLOT]**. Press **ENTER** to select **Plot 1**. Highlight **On** and press **ENTER**. Set **Type** to the normal probability plot which is the third selection in the second row. Press **ENTER**. Set **Data List** to **L1** and **Data Axis** to **X**. For **Marks** select the small square.

Press **ZOOM** and select **9:ZoomStat** and **ENTER**.

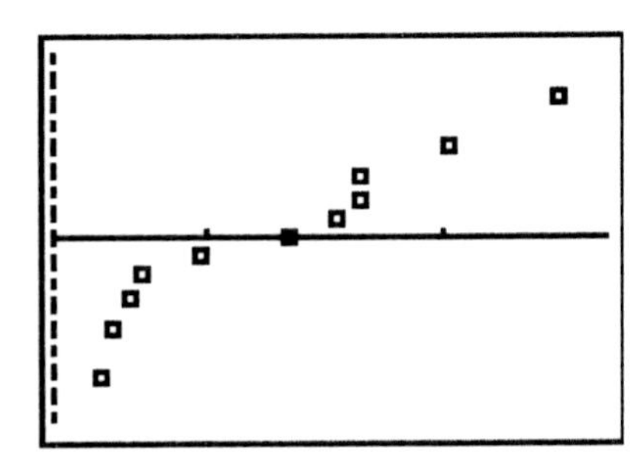

This plot is *fairly* linear, indicating that the data generally follows a normal distribution.

The next step is to find the value of the sample standard deviation. Press **STAT**, highlight **CALC** and select **1:1-Var Stats** and press **ENTER**. Type in **L1** and press **ENTER**. The sample statistics are displayed on the screen. The sample standard deviation, **Sx = 0.6404**.

Next, calculate the test statistic: $\chi^2 = (n-1)s^2 / \sigma_0^2$. The calculation, $10(.6404)^2/(.75)^2 = 7.291$. Because this is a left-tailed test, to find the P-value associated with the test statistic value of 7.291, we calculate the area under the χ^2-curve to the *left* of 7.291. Press **2nd DISTR**, and select **7:** χ^2 **cdf(**. This calculation requires a *lowerbound*, an *upperbound* and the *degrees of freedom.*

The lowerbound is negative infinity (-1E99 on the calculator), the upperbound is 7.291 and the degrees of freedom value is 10 (which is equal to n-1).

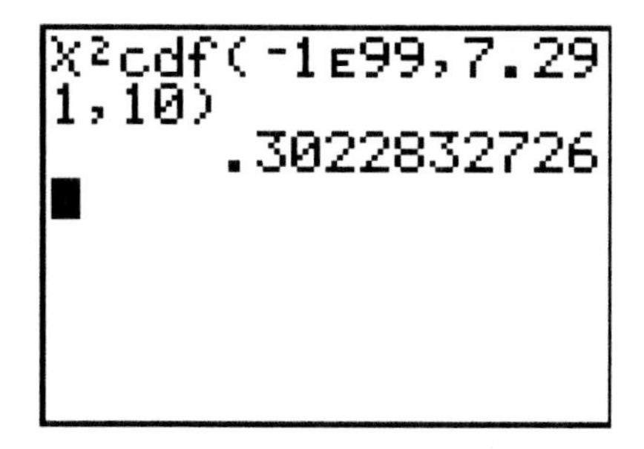

Since the P-value of .3023 is greater than α, the correct conclusion is to **Fail to Reject** H_o. There is not sufficient evidence to support that claim that $\sigma < 0.75$.

◀

▸ Problem 7 (pg. 565) Using the P-value approach

This is a two-tailed hypothesis test about σ. The hypotheses are: H_o: σ = 4.3 vs. H_a: σ ≠ 4.3. The population is normally distributed, the sample size is 12 and the sample standard deviation is 4.8. To run this test, we will use the P-value approach, rather than the Classical approach.

The first step is to calculate the test statistic: $\chi^2 = (n-1)s^2 / \sigma_0^2$. The calculation, $11(4.8)^2/4.3^2 = 13.71$. To find the P-value associated with the test statistic value of 13.71, we calculate the area under the χ^2-curve to the *right* of 13.71. The reason that we are finding the area *to the right* of 13.71 is because the sample standard deviation, 4.8, is *greater than* 4.3. Press **2nd DISTR**, and select **7: χ^2 cdf(.** This calculation requires a *lowerbound*, an *upperbound* and the *degrees of freedom*. The lowerbound is 13.71, the upperbound is positive infinity (1E99 on the calculator) and the degrees of freedom value is 11 (which is equal to n-1).

```
X²cdf(13.71,1E99
,11)
      .2494577162
```

The p-value is .2495. Because this is a two-tailed test, you must compare this P-value to α/2 which is .025. Since the P-value is greater than .025, the correct decision is to **Fail to reject** H_o. There is not sufficient evidence to support that claim that σ ≠ 4.3.

◀

TI-89 Instructions:

These instructions are designed to give you an overview of normal probability plots and one sample hypothesis tests on the TI-89.

Normal Probability Plot
To construct a normal probability plot, first check the y-registers by pressing ◆ and **Y=** and clearing all the y-functions. Return to the **Stats/List Editor**, clear all the lists and enter the data into **list1**. Press **F2** and select **2:Norm Prob Plot**. On the next screen, scroll to the List entry and use 2nd VAR-LINK to select **list1.** Press ENTER. A column labeled **zscores** has been created in the **List Editor.** Press **F2** and select **1:Plot Setup**. On the Plot Setup screen press **F5** to display the normal probability plot.

Hypothesis Test
To conduct a hypothesis test, in the Stats/List Editor press 2nd **F6** and select the appropriate hypothesis test. For **1:ZTest** or **2:TTest,** the next step is to choose the Data Input Method. Choose **Stats** if you are using the summary statistics (sample mean and standard deviation). Choose **Data** if you have entered the actual data values into a list. On the next screen, fill in all the data entry boxes. On the next line, **Alternative Hyp**, use the right arrow and select the correct alternative hypothesis. On the next line, **Results**, use the right arrow and select either **Draw** or **Calculate**. If you select **Draw**, you must clear all plots and y-functions first. Click ENTER until the hypothesis test and summary statistics are displayed.

TI-*n*spire Instructions:

These instructions are designed to give you an overview of normal probability plots and one-sample hypothesis tests on the TI-*n*spire handheld.

Normal Probability Plot

Press (home) and select **6:New Document**. (Note: If you currently have a document open, the next screen will ask if you want to save the document. Press (tab) to select **No**. Press (enter).) Select **3:Add Lists & Spreadsheet. M**ove to the top of column **A** and in the box next to '**A**' type in a name. Move to **Line 1** in column **A** and begin entering your data values.

To construct a normal probability plot, press (home) and select **5: Data & Statistics**. A new page will open with a scatter of points. Press and hold the down arrow to move the cursor to the middle of the x-axis at the bottom of the screen. When the box '**Click to add variable**' appears, press (enter). Highlight the Column **A** name and press (enter). A **Dotplot** of the data in Column A will be displayed. Press (menu), select **1:Plot Type** and select **4: Normal Probability Plot.** Press (ctrl) and the left arrow to return to the Spreadsheet.

Hypothesis Test

To run a hypothesis test, on the Spreadsheet page, press (menu), select **4:Statistics, 4:Stat Tests** and select the appropriate test. For **1:z Test** or **2:t Test**, use the down-arrow to choose the **Data Entry Method**. Choose **Stats** if you are using the summary statistics (sample mean and standard deviation). Choose **Data** if you have entered the actual data values into a list. Press (tab) to highlight **OK** and press (enter). Fill in the entry boxes on the next screen. Press (tab) to move through the entry boxes. (Note: You have the option of plotting the data by using (click) to check the Plot data box. To simply display the numerical output leave the box unselected.) The output will be displayed in two columns in the spreadsheet. To expand a column width, move to any location in that particular column and press (menu), select **1:Actions, 2:Resize** and **1:Resize Column Width.** The column is now highlighted. By clicking the right arrow you can expand the column. Press (enter). Press the down arrow to remove the highlighting.

Inferences on Two Samples

Section 11.1

Example 3 (pg. 588) Testing a Claim Regarding Matched Pairs Data

In this example, the data (found in Table 2 on pg. 586) is paired data, with two reaction times for each of the students. Enter the reaction times for the individual's dominant hand in L1 and enter the reaction times for the individual's non-dominant hand into L2. Next, you must create a set of differences, d = reaction time of dominant hand - reaction time of non-dominant hand. To create this set, move the cursor to highlight the label **L3,** found at the top of the third column, and press ENTER. Notice that the cursor is flashing on the bottom line of the display. Press **2nd [L1]** - **2nd [L2]**

L1	L2	L3 3
.177	.179	------
.21	.202	
.186	.208	
.189	.184	
.198	.215	
.194	.193	
.16	.194	

L3 =L1-L2

and press ENTER.

L1	L2	L3 3
.177	.179	-.002
.21	.202	.008
.186	.208	-.022
.189	.184	.005
.198	.215	-.017
.194	.193	.001
.16	.194	-.034

L3(1)= -.002

Each value in **L3** is the difference **L1 - L2**.

To test the claim that the reaction time in an individual's dominant hand is less than the reaction time in his/her non-dominant hand, the hypothesis test is: $H_o: \mu_d = 0$ vs. $H_a: \mu_d < 0$.

Because the sample size is less than 30, the set of differences must be tested for normality and checked for outliers.

To set up the normal probability plot, press **2nd** [STAT PLOT] . Press ENTER to select **Plot 1**. Highlight **On** and press ENTER. Set **Type** to the normal probability plot which is the third selection in the second row. Press ENTER. Set **Data List** to **L3** and **Data Axis** to **X**. For **Marks,** select the small square.

Press ZOOM and select **9:ZoomStat** and ENTER.

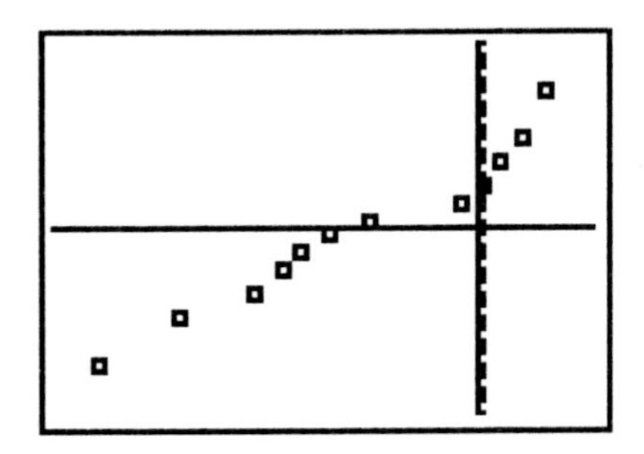

This plot is *fairly* linear, indicating that the data generally follow a normal distribution.

To set up the boxplot, press **2nd [STAT PLOT]**. Press ENTER to select **Plot 1**. Highlight **On** and press ENTER. Set **Type** to the boxplot with outliers which is the first selection in the second row. Press ENTER. Set **XList** to **L3** and **Freq** to **1**. For **Marks** select the small square.

Press ZOOM and select **9:ZoomStat** and ENTER.

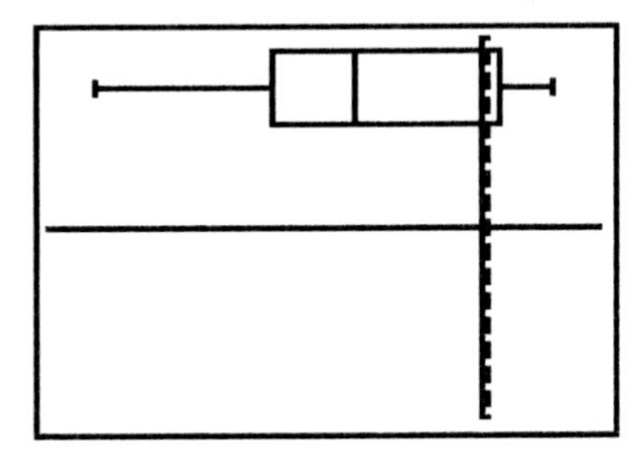

Note: Both of these graphs include the X-axis and Y-axis with the plot. To turn the axes off, press GRAPH . Next press **2nd [FORMAT]** and scroll down to

Axes On. Move the cursor to **Axes Off** and press ENTER . If you redo the graphs, the axes will no longer appear on the screen. You may prefer the way the graphs look without the axes. Here is the graph of the Boxplot:

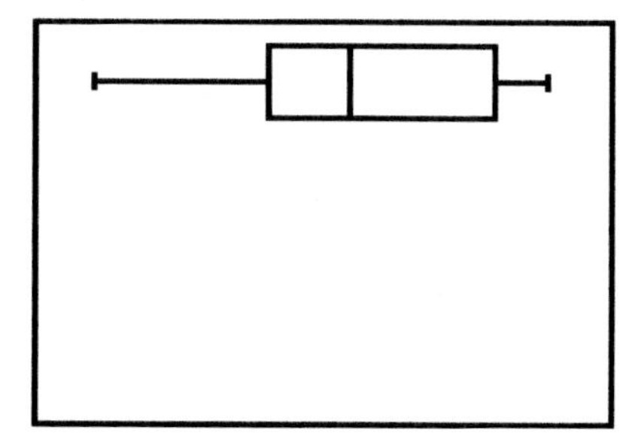

There are no outliers indicated in the boxplot. (Note: Outliers would appear as *'s at the extreme left or right ends of the boxplot.)

To run the hypothesis test, press STAT, highlight **TESTS** and select **2:T-Test**. In this example, you are using the actual data to do the analysis, so select **Data** for **Inpt** and press ENTER. The value for μ_o is **0**, the value in the null hypothesis. The set of differences is found in **L3**, so set **List** to **L3**. Set **Freq** equal to **1**. Choose < μ_o as the alternative hypothesis and highlight **Calculate** and press ENTER.

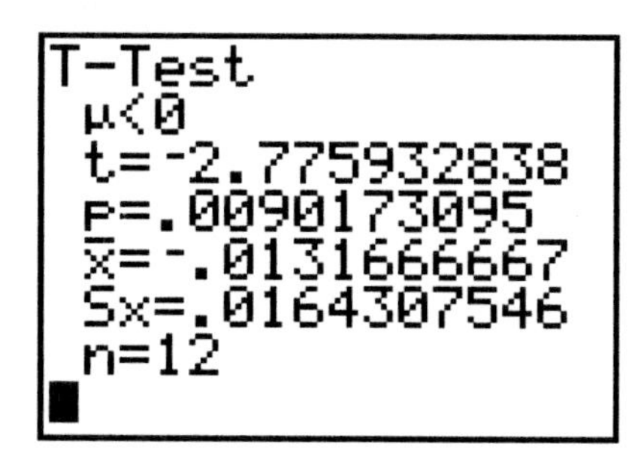

You can also highlight **DRAW** and press ENTER.

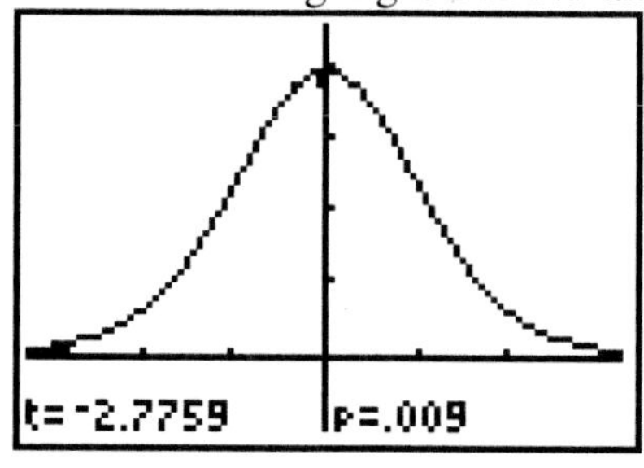

Since the P-value is less than α, the correct decision is to **Reject** H_o. We conclude that an individual's dominant hand has a faster reaction time.

▸ Example 4 (pg. 590) Constructing a Confidence Interval for Matched-Pairs Data

In Example 3 we created a column of differences in L3. We also confirmed that the data was normally distributed with no outliers. To create a 95% confidence interval for the mean difference, press **STAT**, highlight **TESTS** and select **8:T-Interval**. In this example, you are using the actual data to do the analysis, so select **Data** for **Inpt** and press **ENTER**. The set of differences is found in **L3**, so set **List** to **L3**. Set **Freq** equal to **1**. Set **C-Level** to **.95**. Highlight **Calculate** and press **ENTER**.

```
TInterval
 (-.0236,-.0027)
 x̄=-.0131666667
 Sx=.0164307546
 n=12
```

The 95% confidence interval for μ_d, the mean difference in reaction time, is (-.0236, -.0027)seconds. We interpret this interval to mean that, on average, the reaction time of a person's dominant hand is between .0027 and .0236 seconds *faster* than the reaction time of a person's non-dominant hand.

◀

▸ Problem 15 (pg. 591)

In this example the data is paired data, with two measurements of water clarity at the same location in the lake at Joliet Junior College. The first reading is taken at a specific location on the lake at a particular time during a given year. The second reading is taken at the same location 5 years later. Enter the **initial depth** readings into **L1** and the readings **5 years later** into **L2**. Next, you must create a set of differences, d = intial depth reading - reading 5 years later. To create this set, move the cursor to highlight the label **L3,** found at the top of the third column, and press **ENTER**. Notice that the cursor is flashing on the bottom line of the display. Press **2nd [L1]** - **2nd [L2]** and press **ENTER**.

L1	L2	L3 3
38	52	-14
58	60	-2
65	72	-7
74	72	2
56	54	2
36	48	-12
56	58	-2

L3(1)= -14

Each value in **L3** is the difference **L1 - L2**.

We are interesting in testing the claim that the clarity of the lake is improving. If the clarity of the lake *is improving*, then the depth at which the disk is no longer visible should be getting deeper. If that is the case, then the difference, **L1 – L2**, should be *negative*. In other words, the original depths that were measured should be "less deep" than the measurements 5 years later. So the appropriate alternate hypothesis is: $\mu_d < 0$.

Note: A normal probability plot and boxplot of the data indicate that the differences are approximately normal with no outliers.

(b.) To run the hypothesis test, press **STAT**, highlight **TESTS** and select **2:T-Test**. In this example, you are using the actual data to do the analysis, so select **Data** for **Inpt** and press **ENTER**. The value for μ_o is **0,** the value in the null hypothesis. The set of differences is found in **L3**, so set **List** to **L3**. Set **Freq** equal to **1**. Choose < μ_o as the alternative hypothesis and highlight **Calculate** and press **ENTER**.

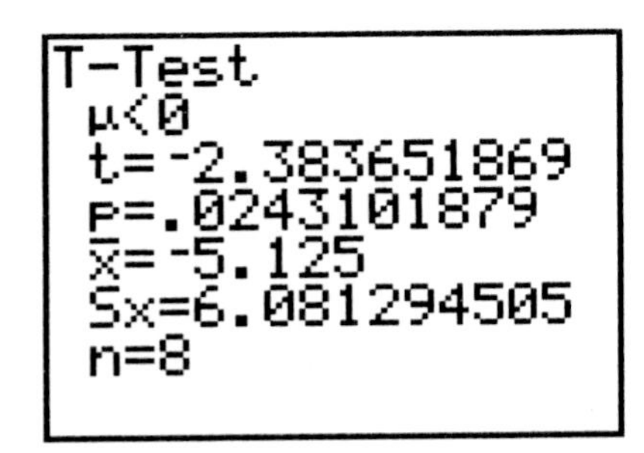

Or, highlight **Draw** and press **ENTER**.

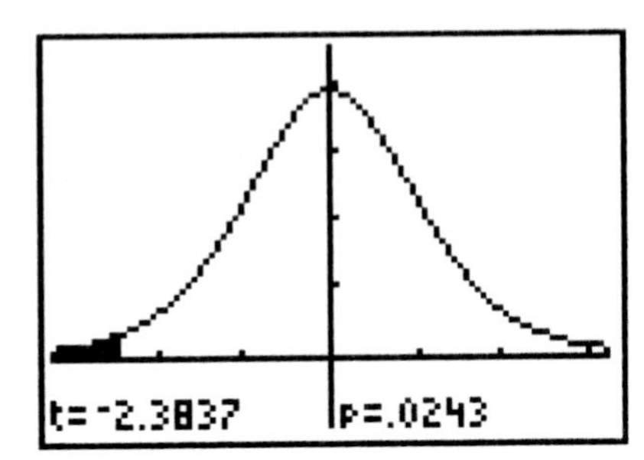

Since the P-value is less than α, the correct decision is to **Reject** H_o. We conclude that the clarity of the lake is improving.

(c) To set up the boxplot, press **2nd [STAT PLOT]**. Press **ENTER** to select **Plot 1**. Highlight **On** and press **ENTER**. Set **Type** to the boxplot with outliers which is the first selection in the second row. Press **ENTER**. Set **XList** to **L3** and **Freq** to **1**. For **Marks** select the small square. Press **ZOOM** and select **9:ZoomStat** and **ENTER**.

▸ Problem 20 (pg. 593)

In this example the data is paired data consisting of daily car rental fees for two different companies at 10 locations in the United States. Enter the rental rates for *Thrifty* into **L1** and the rental rates for *Hertz* into **L2**. Next, create a set of differences, d = *Thrifty* - *Hertz*. Move the cursor to highlight the label **L3,** found at the top of the third column, and press **ENTER**. Press **2nd [L1]** - **2nd [L2]** and press **ENTER**. Each value in **L3** is the difference **L1 - L2**.

We are interested in testing the claim that *Thrifty* is less expensive than *Hertz*. If that is the case, then the difference, **L1 – L2**, should be *negative*. So the appropriate alternate hypothesis is: $\mu_d < 0$.

Note: A normal probability plot and boxplot of the data indicate that the differences are approximately normal with no outliers.

To run the hypothesis test, press **STAT**, highlight **TESTS** and select **2:T-Test**. In this example, you are using the actual data to do the analysis, so select **Data** for **Inpt** and press **ENTER**. The value for μ_o is **0**, the value in the null hypothesis. The set of differences is found in **L3**, so set **List** to **L3**. Set **Freq** equal to **1**. Choose < μ_o as the alternative hypothesis and highlight **Calculate** and press **ENTER**.

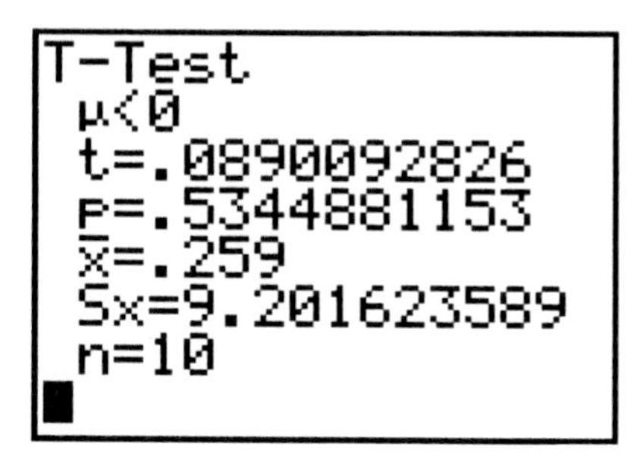

Or, highlight **Draw** and press **ENTER**.

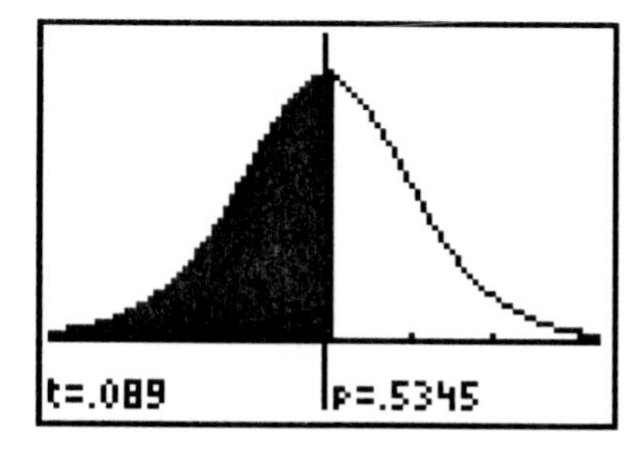

Since the P-value is greater than α, the correct decision is to **Fail to Reject** H_o.
We cannot conclude that *Thrifty* is less expensive than *Hertz*.

◀

Section 11.2

▸ Example 2 (pg. 600) Testing a Claim Regarding Two Means

To test the claim that the flight animals have a different red blood cell mass than the control animals, use a two-tailed test: $H_o: \mu_1 = \mu_2$ vs $H_a: \mu_1 \neq \mu_2$. Refer to Table 3 on pg. 598. Enter the data from the 14 flight rats into **L1** and the data from the control rats into **L2**. Because the data sets are small, the first step is to use normal probability plots and boxplots to verify that both datasets are approximately normal and contain no outliers.

To set up the normal probability plot for the **Flight** data, press **2nd [STAT PLOT]**. Press **ENTER** to select **Plot 1**. Highlight **On** and press **ENTER**. Set **Type** to the normal probability plot which is the third selection in the second row. Press **ENTER**. Set **Data List** to **L1** and **Data Axis** to **X**. For **Marks,** select the small square. Press **ZOOM** and select **9:ZoomStat** and **ENTER**.

To set up the normal probability plot for the **Control** data, repeat the above steps changing the **Data List** to **L2**.

To set up both boxplots on the same graph, press **2nd** [STAT PLOT] . Press **ENTER** to select **Plot 1**. Highlight **On** and press **ENTER**. Set **Type** to the boxplot with outliers which is the first selection in the second row. Press **ENTER**. Set **XList** to **L1** and **Freq** to **1**. For **Marks** select the small square. Use the up arrow to move to the top of the screen and highlight **Plot 2**. Highlight **On** and press **ENTER**. Set **Type** to the boxplot with outliers which is the first selection in the second row. Press **ENTER**. Set **XList** to **L2** and **Freq** to **1**. For **Marks** select the small square.

Press **ZOOM** and select **9:ZoomStat** and **ENTER**.

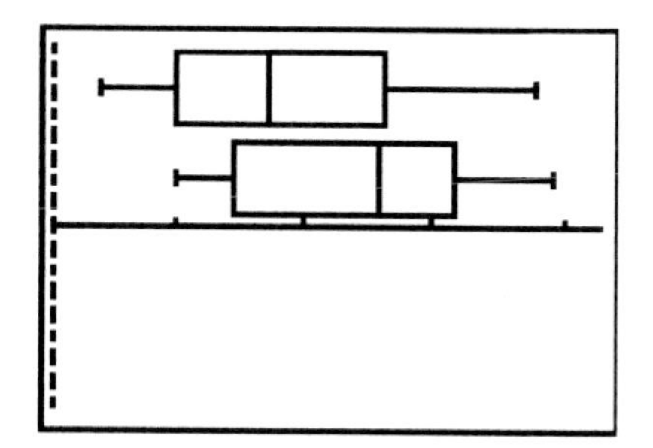

By drawing both boxplots on the same graph, you can visually compare the two datasets. Notice that the Flight data graph appears shifted slightly to the left of the Control data graph. The next step is to determine whether this shift is statistically significant.

To run the hypothesis test, press STAT, highlight **TESTS**, and select **4:2-SampTTest**. Since you are inputting the sample data, select **Data** and press ENTER. Enter **L1** for **List1** and **L2** for **List 2**. Set **Freq1** and **Freq2** to **1**. Select $\neq \mu_2$ as the alternative hypothesis and press ENTER. Scroll down to the next line. On this line, there are two options. Select **NO** because, in the procedure we are using (called Welch's approximate t-test), the variances are NOT assumed to be equal and therefore, we do not want a pooled variance. Press ENTER.

```
2-SampTTest
 Inpt:Data Stats
 List1:L1
 List2:L2
 Freq1:1
 Freq2:1
 μ1:≠μ2 <μ2 >μ2
↓Pooled:No Yes
```

Scroll down to the next line, highlight **Calculate** and press ENTER.

```
2-SampTTest
 μ1≠μ2
 t=-1.436781704
 p=.1627070766
 df=25.99635232
 x̄1=7.880714286
↓x̄2=8.43
```

The output (shown above) for **Calculate** displays the alternative hypothesis, the test statistic, the P-value, the degrees of freedom and the sample statistics. Notice the degrees of freedom = 25.996. In cases, such as this one, in which the population variances are not assumed to be equal, the calculator calculates an adjusted degrees of freedom, (see the formula on pg. 599 of your textbook.)

If you choose **Draw**, the output includes a graph with the area associated with the P-value shaded.

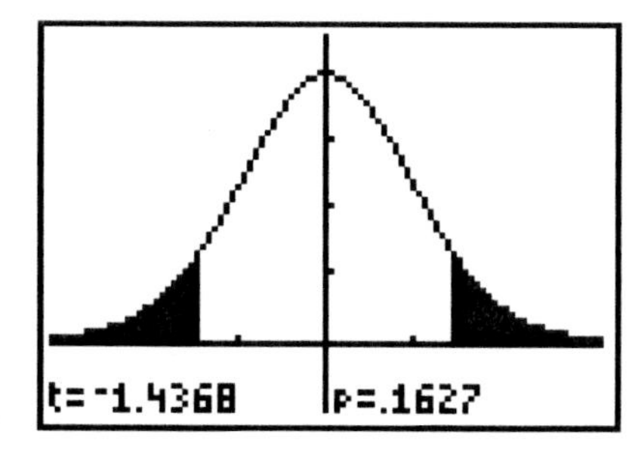

Since the P-value is greater than α, we fail to reject H_0. There is not sufficient evidence to support the claim that there is a significant difference in the red blood cell mass of the flight animals and the control animals.

◀

▸ Example 3 (pg.602) Constructing a Confidence Interval about the Difference of Two Means

For this example, we will use the summary data from Table 3 on pg. 598. In Example 2, we tested for normality using normal probability plots and for outliers using Boxplots. Both data sets appeared to be approximately normal with no outliers.

To construct a 95% confidence interval for (μ_1-μ_2), press **STAT**, highlight **TESTS** and select **0:2-SampTint**. For **Inpt,** select **Stats** and press **ENTER**. Enter the mean of the first sample, 7.881, the standard deviation, 1.017, and the sample size, 14. Then enter the mean (8.43), standard deviation (1.005), and sample size (14) of the second sample. Scroll down to the next line and enter .95 for the **C-level**. On the next line, select **No**, because we are not using a pooled variance.

```
2-SampTInt
 Inpt:Data Stats
 x̄1:7.881
 Sx1:1.017
 n1:14
 x̄2:8.43
 Sx2:1.005
↓n2:14
```

Scroll down to the next line, **Calculate,** and press **ENTER**

```
2-SampTInt
 (-1.334,.23648)
 df=25.9963378
 x̄1=7.881
 x̄2=8.43
 Sx1=1.017
↓Sx2=1.005
```

A 95 % confidence interval for the difference in the population means is (-1.334, .23648). Since this interval contains 0, the correct conclusion is that there is no difference in the red blood cell mass of the two groups. (Note: This interval differs slightly from the textbook interval because the calculator uses the adjusted degrees of freedom in the calculations.)

◀

▸ Problem 10 (pg. 604)

To test the claim that the treatment group experienced a larger mean improvement than the control group, use a one-tailed test: H_o: $\mu_1 = \mu_2$ vs H_a: $\mu_1 > \mu_2$. Because the data sets are large (n_1 and $n_2 > 30$), we know that the sampling distribution of ($\bar{x}_1 - \bar{x}_2$) is approximately normal and we can safely use the Two-sample T-test.

To run the hypothesis test, press **STAT**, highlight **TESTS**, and select **4:2-SampTTest**. Since you are inputting the sample statistics, select **Stats** and press **ENTER**. Enter the sample statistics for the two samples.

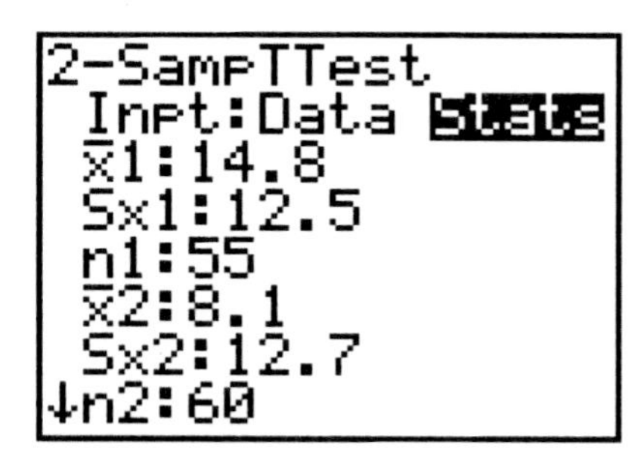

Scroll down to the next line, select $>\mu_2$ as the alternative hypothesis and press **ENTER**. Scroll down to the next line. Select **NO** because we are not using a pooled variance. Press **ENTER**.

Scroll down to the next line, highlight **Calculate** and press **ENTER**.

```
2-SampTTest
 μ1≠μ2
 t=-1.436781704
 p=.1627070766
 df=25.99635232
 x̄1=7.880714286
↓x̄2=8.43
```

The output (shown above) for **Calculate** displays the alternative hypothesis, the test statistic, the P-value, the adjusted degrees of freedom and the sample statistics.

Or, select **Draw** and press **ENTER**.

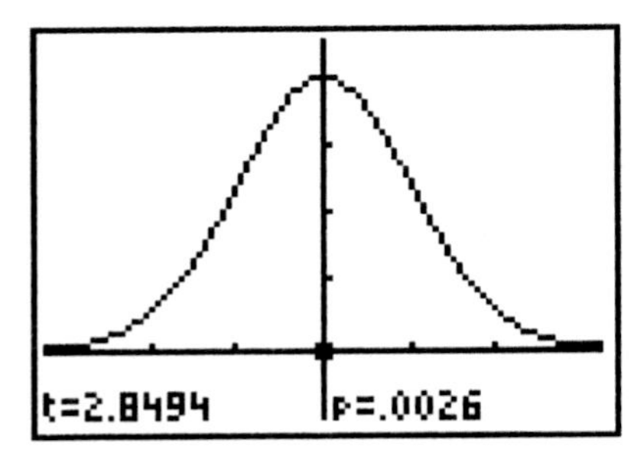

Since the P-value is less than α, we reject H_0. The data supports the claim that the treatment group has a larger mean improvement than the control group.

◀

Problem 15 (pg. 605)

To test the claim that carpeted rooms contain more bacteria than uncarpeted rooms, use a one-tailed test: H_o: $\mu_1 = \mu_2$ vs H_a: $\mu_1 > \mu_2$. Enter the data from the 8 carpeted rooms into **L1** and the data from the uncarpeted rooms into **L2**. A normal probability plot and boxplot indicate that the data are normally distributed with no outliers.

To run the hypothesis test, press **STAT**, highlight **TESTS**, and select **4:2-SampTTest**. Since you are inputting the sample data, select **Data** and press **ENTER**. Enter **L1** for **List1** and **L2** for **List 2**. Set **Freq1** and **Freq2** to **1**. Select $>\mu_2$ as the alternative hypothesis and press **ENTER**. Scroll down to the next line. Select **NO** because, in the procedure we are using, we do not want a pooled variance. Press **ENTER**. Scroll down to the next line, highlight **Calculate** and press **ENTER**.

```
2-SampTTest
 μ1>μ2
 t=.9557706856
 p=.1779571293
 df=13.56321553
 x̄1=11.2
↓x̄2=9.7875
```

The output (shown above) for **Calculate** displays the alternative hypothesis, the test statistic, the P-value, the adjusted degrees of freedom and the sample statistics.

If you choose **Draw**, the output includes a graph with the area associated with the P-value shaded.

Since the P-value (.178) is greater than α, we fail to reject H_0. There is not sufficient evidence to support the claim that there is more bacteria in carpeted rooms than in uncarpeted rooms.

Section 11.3

▸ Example 2 (pg. 612) Testing a Claim Regarding Two Population Proportions

Refer to Example 1 on pg. 610. To test the claim that the proportion of Nasonex users who experienced headaches as a side effect is greater than the proportion in the control group who experienced headaches, the correct hypothesis test is: $H_o: p_1 = p_2$ vs $H_a: p_1 > p_2$. Designate the Nasonex users as Group 1 and the Control Group as Group 2. The sample statistics are $n_1 = 2103$, $x_1 = 547$, $n_2 = 1671$, and $x_2 = 368$.

First, verify that the requirements for the hypothesis test are satisfied. The problem states that the individuals were randomly divided into two groups so the first requirement (independent random samples) is satisfied. The second requirement is: $n\hat{p}(1-\hat{p}) \geq 10$ for each of the groups. For the first group, $\hat{p} = \frac{x}{n} = \frac{547}{2103} = 0.26$. The calculation, $n\hat{p}(1-\hat{p}) = 2103 * 0.26 * (1-0.26)$ is greater than 10. Repeat this calculation for the second group. Both calculations are greater than 10 so the second requirement is satisfied. The final requirement is that the sample sizes are not more than 5% of the population sizes. The population of Americans 12 years of age or older is in the millions so this requirement is easily satisfied.

Next, to run the hypothesis test, press **STAT**, highlight **TESTS** and select **6:2-PropZTest** and fill in the appropriate information. Highlight **Calculate** and press **ENTER**.

```
2-PropZTest
 p1>p2
 z=2.839330068
 p=.0022604818
 p̂1=.2601046125
 p̂2=.2202274087
↓p̂=.2424483307
```

The output displays the alternative hypothesis, the test statistic, the P-value, the sample proportions, the weighted estimate of the population proportion, $\hat{p}$, and the sample sizes.

Or, highlight **Draw** and press ENTER.

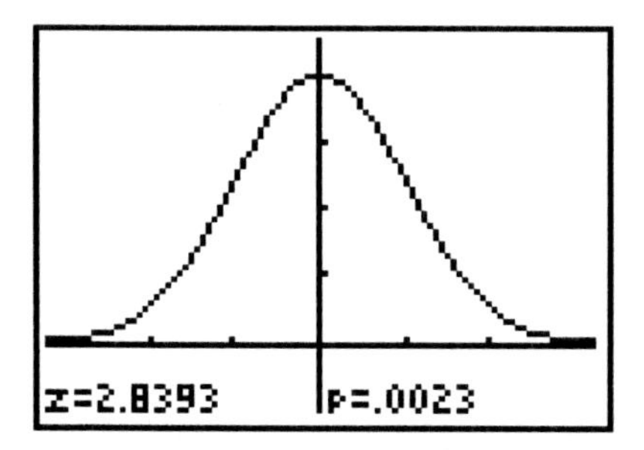

Since the P-value is less than α, the correct decision is to **Reject** H_o. There is sufficient evidence to support the claim that there is a higher incidence of headaches among the Nasonex users than among the individuals taking a placebo.

Example 3 (pg. 613) A Confidence Interval for the Difference Between Two Population Proportions

Construct a confidence interval to compare American views about the moral values of the country in 2002 vs. 2007. Designate the Americans in the 2007 sample as Group 1 and the Americans in the 2002 sample as Group 2. The sample statistics are $n_1 = 1100$, $x_1 = 902$, $n_2 = 1100$, and $x_2 = 737$.

First, verify that the requirements for the hypothesis test are satisfied. They were obtained independently so the first requirement (independent random samples) is satisfied. The second requirement is: $n\hat{p}(1-\hat{p}) \geq 10$ for each of the groups. For the first group, $\hat{p} = \frac{x}{n} = \frac{902}{1100} = 0.82$. The calculation, $n\hat{p}(1-\hat{p}) = 1100 * 0.82 * (1-0.82)$ is greater than 10. Repeat this calculation for the second group. Both calculations are greater than 10 so the second requirement is satisfied. The final requirement is that the sample sizes are not more than 5% of the population sizes. The population of Americans is in the millions so this requirement is easily satisfied.

Press **STAT**, highlight **TESTS** and select **B:2-PropZInt** and fill in the appropriate information. Highlight **Calculate** and press **ENTER**.

```
2-PropZInt
 (.11989,.18011)
 p̂1=.82
 p̂2=.67
 n1=1100
 n2=1100
```

The confidence interval is (.120, .180). The proportion of Americans who believe that the state of moral values in the country is getting worse increased between 12% and 18% from 2002 to 2007.

Section 11.4

Example 2 (pg. 625) Testing a Claim Regarding Two Population Standard Deviations

To test the claim that Cisco Systems is a more volatile stock than General Electric, use a one-tailed test: H_o: $\sigma_1 = \sigma_2$ vs H_a: $\sigma_1 > \sigma_2$. Use the data found in Table 6 on pg. 626. Enter the data from the Cisco Systems stock into **L1** and the data from the General Electric stock into **L2**.

The hypothesis test we are using requires that the data be normally distributed. Even minor deviations from normality will affect the validity of the test. (Note: In other hypothesis tests, minor deviations from normality did not seriously affect the test validity.) So, an important first step is to use normal probability plots to verify that both datasets are normally distributed. Once this has been confirmed, we can run the hypothesis test.

To run the hypothesis test, press **STAT**, highlight **TESTS**, and select **D:2-SampFTest**. Since you are inputting the sample data, select **Data** and press **ENTER**. Enter **L1** for **List1** and **L2** for **List 2**. Set **Freq1** and **Freq2** to **1**. On the next line, select $>\sigma_2$ as the alternative hypothesis. Highlight **Calculate** and press **ENTER**.

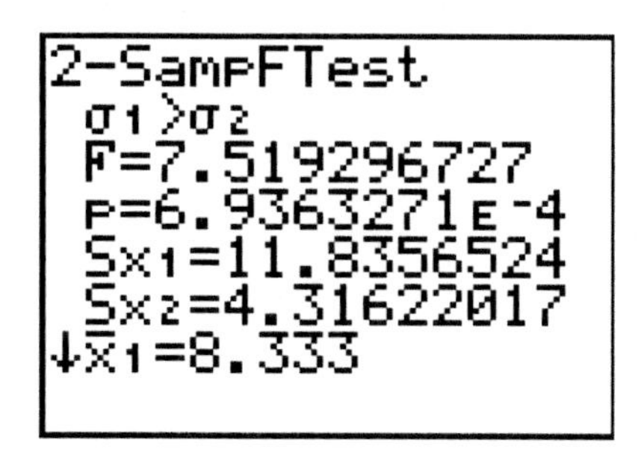

The output displays the alternative hypothesis, the test statistic (F), the P-value, the sample standard deviations, the sample means and the sample sizes.

Or, highlight **Draw** and press **ENTER**.

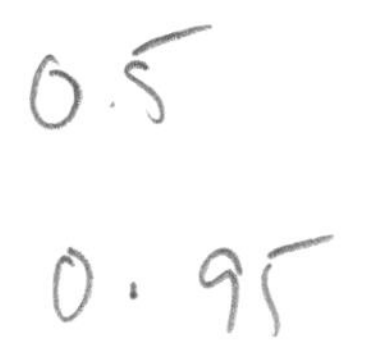

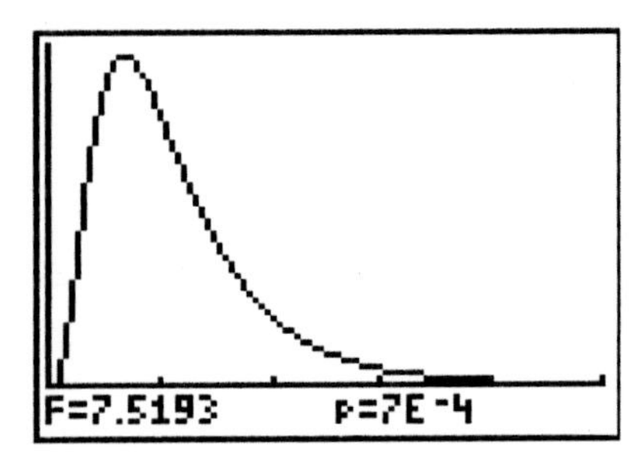

The output shows the F-distribution curve, the value of the F-statistic and the P-value. Since the P-value (.0007) is less than α, the correct decision is to **Reject** H_o. There is sufficient evidence to support the claim that Cisco Systems stock is more volatile than General Electric stock.

▸ Example 3 (pg. 627) Testing a Claim Regarding Two Population Standard Deviations

To test the claim that the standard deviation of the red blood cell count in the flight animals is different from the standard deviation of the red blood cell count in the control animals , use a two-tailed test: H_o: $\sigma_1 = \sigma_2$ vs. H_a: $\sigma_1 \neq \sigma_2$. Enter the data from the flight rats into **L1** and the data from the control rats into **L2**. Use normal probability plots to verify that each data set is normally distributed.

To run the hypothesis test, press **STAT**, highlight **TESTS**, and select **D:2-SampFTest**. Since you are inputting the sample data, select **Data** and press **ENTER**. Enter **L1** for **List1** and **L2** for **List 2**. Set **Freq1** and **Freq2** to **1**. On the next line, select $\neq \sigma_2$ as the alternative hypothesis. Highlight **Calculate** and press **ENTER**.

```
2-SampFTest
 σ1≠σ2
 F=1.023974926
 p=.9665798398
 Sx1=1.0174513
 Sx2=1.00546966
↓x̄1=7.880714286
```

The output displays the alternative hypothesis, the test statistic (F), the P-value, the sample standard deviations, the sample means and the sample sizes.

Or, highlight **Draw** and press **ENTER**.

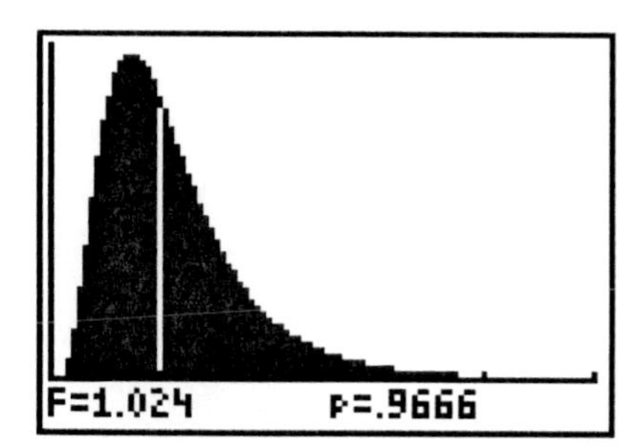

The output shows the F-distribution curve, the value of the F-statistic and the P-value. Since the P-value (.9666) is greater than α, the correct decision is to **Fail to Reject** H_o. There is not sufficient evidence to support the claim that the standard deviation of the red blood cell count for the flight rats is different from the standard deviation of the red blood cell count for the control group.

TI-89 Instructions:

These instructions are designed to give you an overview of two sample confidence intervals and hypothesis tests on the TI-89.

To create a confidence interval, in the Stats/List Editor press 2nd **F7** and select the appropriate interval:

If the data is matched pairs, select **2:TInterval.** The next step is to choose the Data Input Method. Choose **Stats** if you are using the summary statistics (sample mean and standard deviation). Choose **Data** if you have entered the actual data values into a list. On the next screen, fill in all the information and click ENTER until the confidence interval and summary statistics are displayed.

If you are using two independent samples and you are constructing a confidence interval for the difference in population means, choose either **3:2-SampZInt** (if the population standard deviations are given), or **4:2-SampTInt** (if you are using the sample standard deviations). On the next screen, select the Input Method. Fill in all the entry boxes. If you are using a pooled standard deviation for the **2-SampTInt** select **Yes** for **Pool**. Otherwise select **No**. Click ENTER until the confidence interval and summary statistics are displayed.

To construct a confidence interval for the difference in two population proportions, select **6:2-PropZInt.**

To conduct a hypothesis test, in the Stats/List Editor press 2nd **F6** and select the appropriate hypothesis test: **2:TTest, 3:2-SampZTest, 4:2-SampTTest** or **6:2-PropZInt.**

TI-*n*spire Instructions:

These instructions are designed to give you an overview of two sample confidence intervals and hypothesis tests on the TI-*n*spire handheld.

Press (home) and select **6:New Document**. (Note: If you currently have a document open, the next screen will ask if you want to save the document. Press (tab) to select **No**. Press (enter).) If you are working with the actual data, select **3:Add Lists & Spreadsheet** and enter the data into the spreadsheet. If you are working with the summary statistics, select **1:Add Calculator.**

To create a confidence interval, press (menu), select **Statistics, Confidence Intervals** and select the appropriate interval.

If the data is matched pairs, select **2:TInterval.** The next step is to choose the Data Input Method. Choose **Stats** if you are using the summary statistics (sample mean and standard deviation). Choose **Data** if you have entered the actual data values into a list. Press (tab) to highlight **OK** and press (enter). On the next screen, fill in all the information and press (tab) to highlight **OK** and press (enter). The confidence interval is displayed along with the summary statistics.

If you are using two independent samples and you are constructing a confidence interval for the difference in population means, choose either **2-SampZInt** (if the population standard deviations are given), or **2-SampTInt** (if you are using the sample standard deviations). On the next screen, select the Input Method. Use (tab) and (enter) to fill in all the entry boxes. If you are using a pooled standard deviation in the **2-SampTInt** select **Yes** for **Pool**. Otherwise select **No**. Press (tab) to highlight **OK** and press (enter). The confidence interval is displayed along with the summary statistics.

To construct a confidence interval for the difference in two population proportions, select **2-PropZInt.**

To conduct a hypothesis test, select the appropriate hypothesis test: **TTest, 2-SampZTest, 2-SampTTest** or **2-PropZInt.**

Note: To clear the output on the Calculator page, highlight the output and press (clear).

Inference on Categorical Data

Section 12.1

▸ Example 3 (pg. 649) Testing a Claim using the Goodness-of-Fit Test

In this example, we test the claim that the population distribution of the United States is different now than it was in 2000. The procedure that we use is the Chi-Square (χ^2) Goodness-of-Fit test.

The χ^2 test has 3 requirements: (1) the data are randomly selected; (2) all *expected frequencies* are greater than or equal to 1 and (3) no more than 20% of the *expected frequencies* are less than 5. For the first requirement, we *assume* that the data was randomly selected. For requirements (2) and (3), check the data and make sure that these requirements are satisfied.

Using Table 2 on pg. 648, enter the observed values into **L1** and the expected values into **L2**.

L1	L2	L3 2
274	285	------
303	343.5	
564	534	
359	337.5	

L2(5) =

This test is a test of the hypotheses: H_o: The distribution of residents in the U.S. is the same today as it was in 2000, vs. H_a: The distribution of residents in the U.S. is different today than in 2000.

To perform the goodness of fit test, press **STAT**, highlight **TESTS**, and select **D: χ^2GOF-Test** and press **ENTER**.

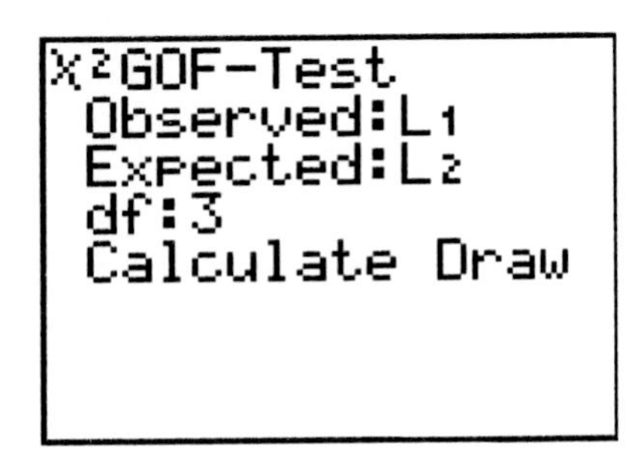

For **Observed**, **L1** should be selected. For **Expected**, **L2** should be selected. Move the cursor to the next line, and set **df** to 'the number of categories minus 1.' Select **Calculate** and press ENTER.

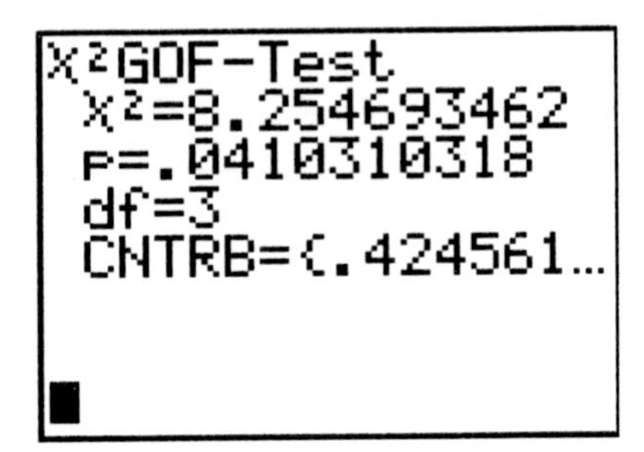

The output displays the χ^2 statistic, the P-value and degrees of freedom. Since the P-value (.041) is less than α (.05), the correct decision is to **reject** the null hypothesis. There is sufficient evidence to support the claim that the distribution of residents in the U.S. is different today than it was in 2000.

Notice the last row of information in the output. This is a partial list of the contributions of each individual cell to the total χ^2 value. To see the entire list, press STAT and select **1:Edit**. Move the cursor so that it is highlighting **L3** at the top of the third column and press ENTER. Press 2nd [**List**] and select **CNTRB.** Press ENTER and the list will be stored in L3.

L1	L2	L3 3
274	285	[illegible]
303	343.5	4.7751
564	534	1.6854
359	337.5	1.3696
------	------	------

L3(1)=.4245614035...

▸ Example 4 (pg. 650) Testing a Claim Using the Goodness Of Fit Test

In this example, we test the claim that the day (Sunday, Monday, Tuesday, etc.) on which a child is born occurs with equal frequency. This is a test of the hypotheses: H_o: $p_1 = p_2 = p_3 = p_4 = p_5 = p_6 = p_7 = \frac{1}{7}$ vs. H_a: At least one of the proportions is different than the others.

Enter the fraction $\frac{1}{7}$ (by pressing 1 ÷ 7) into **L1** seven times (for the seven days of the week). Move the cursor so that it is flashing on '**L2**' at the top of the second column and press **ENTER**. The cursor will move to the bottom of the screen and will be flashing next to '**L2=**'. Type in **L1*500**. (Note: 500 is the sample size.) Press **ENTER**. **L2** will contain the *expected frequencies.*

Enter the *observed frequencies* into **L3**.

L1	L2	L3 3
.14286	71.429	78
.14286	71.429	74
.14286	71.429	76
.14286	71.429	71
.14286	71.429	81
.14286	71.429	63
------	------	

L3(8) =

Notice the *expected frequencies* in L2. All the values are greater than 5, so the requirements of the test have been satisfied.

To perform the goodness of fit test, press **STAT**, highlight **TESTS**, and select **D: χ^2GOF-Test** and press **ENTER**. For **Observed, L3** should be selected. For **Expected, L2** should be selected. Move the cursor to the next line, and set **df** to 'the number of categories minus 1.' Select **Calculate** and press **ENTER**.

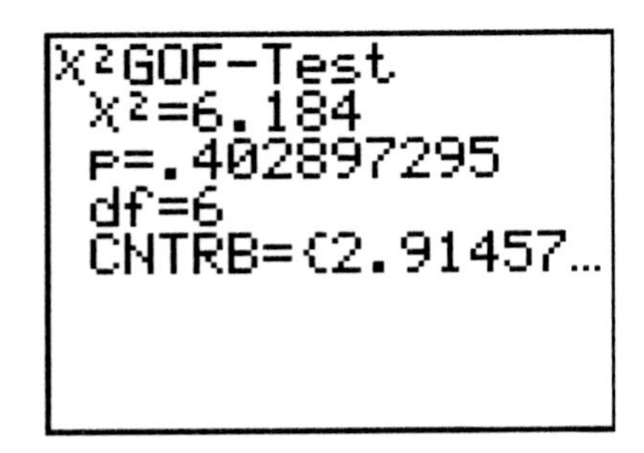

The output displays the χ^2 statistic, the P-value and degrees of freedom. Since the P-value (.403) is greater than α (.05), the correct decision is to **fail to reject** the null hypothesis. There is not sufficient evidence to reject the belief that the day of the week on which a child is born occurs with equal frequency.

◀

▸ Problem 25 (pg.655)

(a.) In this problem, we will use the TI-84 to generate 500 random integers numbered 1 through 5. The first step is to set the *seed* by selecting a 'starting number' and storing this number in **rand**. Suppose, for this example, that we select the number '22' as the starting number. Type **22** into your calculator and press the **STO** key. Next press the **MATH** key and select **PRB** and select **rand,** which stands for 'random number'. Press **ENTER** and the starting value of '22' will be stored into **rand** and will be used as the *seed* for generating random numbers.

Now you are ready to generate the random integers. Press **MATH** and select **PRB**. Select **5:RandInt(.** This function requires three values: the starting value, the ending value and the number of values you want to generate. For this example, you want to generate 500 values from the integers ranging from 1 to 5. The command is **randInt(1,5,500)**. Press **STO** **L1** and press **ENTER**. The values will be stored into **L1**.

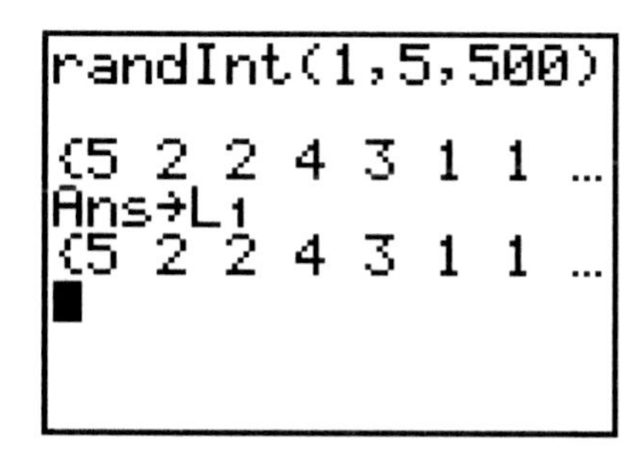

Next, press **2nd** **[LIST]**. Select **OPS** and **1:SortA(** . Type in **L1** and press **ENTER**. This will put all the values in **L1** in numerical order. Press **2nd** **[QUIT]**. Press **STAT** , select **EDIT.** Move the cursor to the first entry in L1 and scroll through the list using the down arrow. Hold the 'down arrow' button down and scroll through the 1's. Record the number of 1's that you have on a piece of paper.

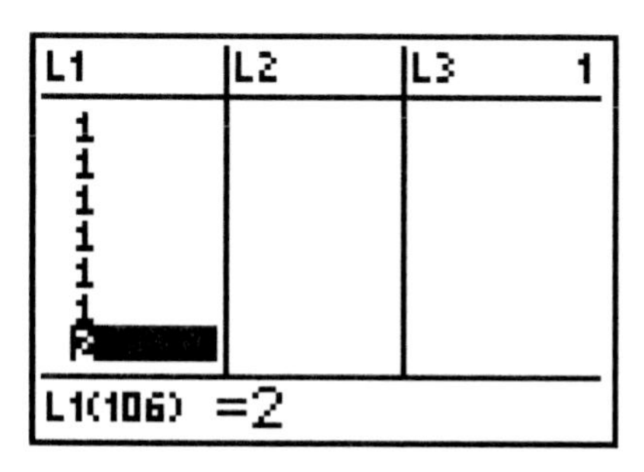

In this illustration, there are 105 1's. (Note: If you started with a different seed, you would get a different result).

Continue scrolling through the column. Stop at the final '2' in the list.

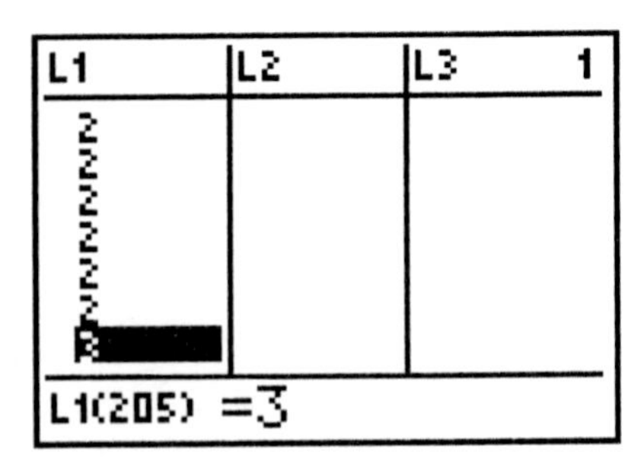

The number of 2's in this illustration is 204 – 105 = 99. Record this number on your paper.

Continue this process until you have recorded the counts for all the data values in L1.

(b.) Each of the numbers, 1 through 5, should occur with equal frequency. So, the proportion of 1's, 2's, 3's, 4's and 5's should equal .20.

(c.) In this example, we are testing the claim that the random number generator is generating random numbers between 1 and 5. This is a test of the hypotheses: H_o: $p_1 = p_2 = p_3 = p_4 = p_5 = .2$, vs. H_a: At least one of the proportions is different than the others.

Press **STAT**, select **EDIT. L1** contains the 500 randomly selected integers. Enter the values 1 through 5 into **L2** and their corresponding frequencies into **L3**. L3 contains the *observed frequencies*

L1	L2	L3 3
1	1	105
1	2	98
1	3	105
1	4	88
1	5	103
1	------	
1		

L3(6) =

Move to **L4** and enter the expected proportion, .20, five times. Move the cursor so that it is flashing on 'L5' at the top of the fifth column and press **ENTER**. The cursor will move to the bottom of the screen and will be flashing next to '**L5=**'. Type in **L4*500**. (Note: 500 is the sample size.) Press **ENTER**. **L5** will contain the *expected frequencies.* All the *expected frequencies* in L5 are greater than 5, so the requirements of the test have been satisfied.

To perform the goodness of fit test, press **STAT**, highlight **TESTS**, and select **D: χ^2GOF-Test** and press **ENTER**.

For **Observed**, **L3** should be selected. For **Expected**, **L5** should be selected. Move the cursor to the next line, and set **df** to 'the number of categories minus 1.' Select **Calculate** and press **ENTER**.

The output displays the χ^2 statistic, the P-value and degrees of freedom. Since the P-value (.723) is greater than α (.05), the correct decision is to **fail to reject** the null hypothesis. There is not sufficient evidence to reject the claim that the random number generator is generating random numbers between 1 and 5. In other words, we can conclude that the random number generator is working correctly.

◀

Section 12.2

Example 2 (pg. 660) Performing a Chi-Square Independence Test

Use the data in Table 10 on pg. 661 to test the claim that marital status and happiness level are independent. The correct procedure is the Chi-Square (χ^2) Independence test.
The hypotheses are: H_o : Marital status and Happiness level are independent (or not related), vs. H_a : Marital status and Happiness level are dependent (or somehow related).

The χ^2 test has 3 requirements: (1) the data are randomly selected; (2) all *expected frequencies* are greater than or equal to 1 and (3) no more than 20% of the *expected frequencies* are less than 5. For the first requirement, we *assume* that the data was randomly selected. We will verify requirements (2) and (3) at the end of our analysis.

The first step is to enter the data in the table into **Matrix A**. On the TI-84 Plus and TI-83 Plus, press **2nd [MATRIX]**. **MATRIX** is found above the $\mathbf{x^{-1}}$ key. (On the TI-83, press **MATRX**). Highlight **EDIT** and press **ENTER**.

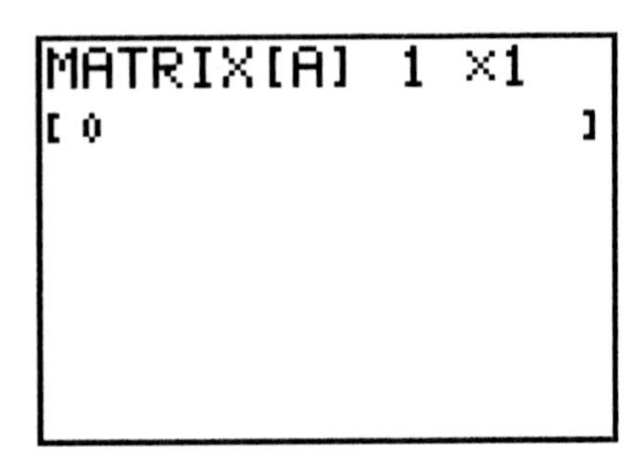

On the top row of the display, enter the size of the matrix. The matrix has 3 rows and 4 columns, so press **3**, press the right arrow key, and press **4**. Press **ENTER**. Enter the first value, **600**, and press **ENTER**. Enter the second value, **63**, and press **ENTER**. Continue this process and fill the matrix.

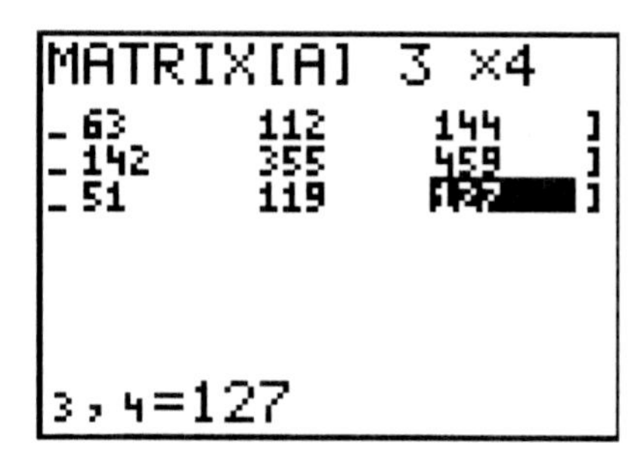

Press **2nd [Quit]**. To perform the test of independence, press **STAT**, highlight **TESTS**, and select **C: χ^2-Test** and press **ENTER**.

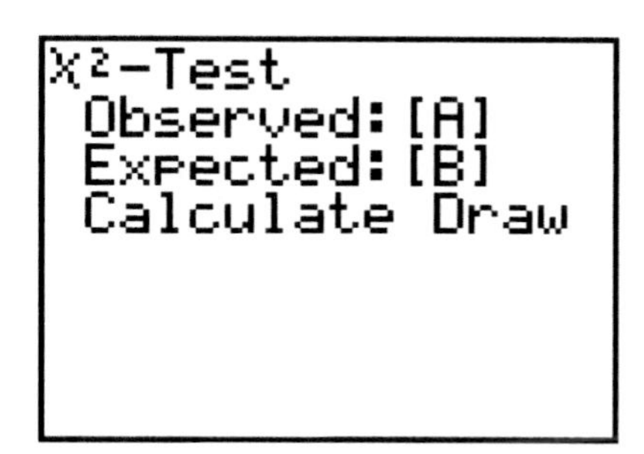

For **Observed**, **[A]** should be selected. If **[A]** is not already selected, press **2nd [MATRX]**, highlight **NAMES**, select **1:[A]** and press **ENTER**. For, **Expected**, **[B]** should be selected. Move the cursor to the next line and select **Calculate** and press **ENTER**.

χ²-Test
χ²=224.1155847
p=1.378872E-45
df=6

The output displays the test statistic and the P-value. Since the P-value is less than α, the correct decision is to **Reject** the null hypothesis. This means that Marital status and Happiness level are dependent. We conclude that marital status and happiness level are related.

Or (after clearing all graphs and Y-registers), you could highlight **Draw** and press **ENTER**.

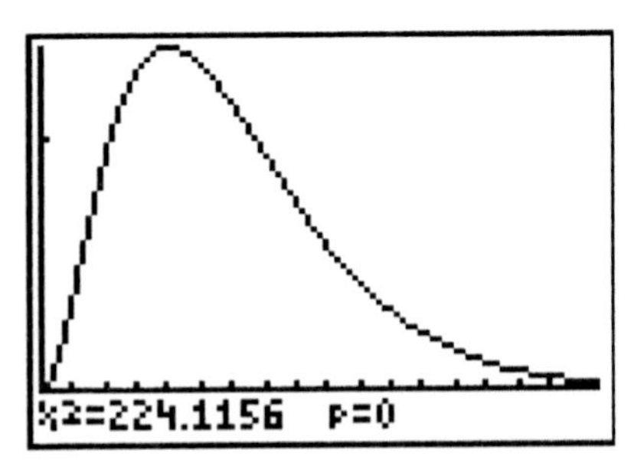

This output displays the χ^2 **–curve** with the area associated with the P-value shaded in. The test statistic and the P-value are also displayed.

The final step in this procedure is to confirm that the test requirements have been satisfied. The two requirements that we need to verify are: (1) all *expected frequencies* are greater than or equal to 1 and (2) no more than 20% of the *expected frequencies* are less than 5. Both requirements involve the *expected frequencies* which are stored in **Matrix B**. To view **Matrix B,** press **2nd [MATRX]** highlight **NAMES**, select **2:[B]** and press **ENTER** **ENTER**.

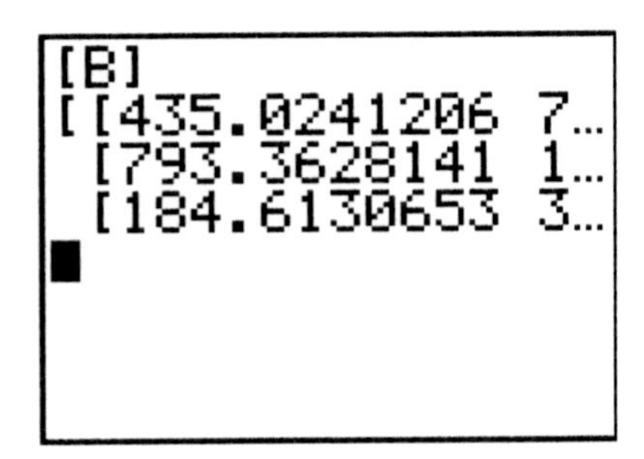

Scroll through the 8 entries in **Matrix B** and confirm that all entries are greater than 5. This confirms that the test requirements have been satisfied.

◀

Example 5 (pg. 664) A Test of Homogeneity for Proportions

In this example, we are testing the claim that the proportions of subjects who experience abdominal pain are equal among all three groups (those taking Zocor, those taking a placebo and those taking Cholestyramine). The correct procedure is the Chi-Square (χ^2) Independence test. The hypotheses are: H_o: $p_1 = p_2 = p_3$ vs. H_a: At least one of the proportions is different than the others.

The χ^2 test has 3 requirements: (1) the data are randomly selected; (2) all *expected frequencies* are greater than or equal to 1 and (3) no more than 20% of the *expected frequencies* are less than 5. For the first requirement, we *assume* that the data was randomly selected. We will verify requirements (2) and (3) at the end of our analysis.

The first step is to enter the data in the table into **Matrix A**. On the TI-84 Plus and TI-83 Plus, press **2nd [MATRIX]**. **MATRIX** is found above the **x^{-1}** key. (On the TI-83, press **MATRX**). Highlight **EDIT** and press **ENTER**.

On the top row of the display, enter the size of the matrix. The matrix has 2 rows and 3 columns, so press **2**, press the right arrow key, and press **3**. Press **ENTER**. Enter the first value, **51**, and press **ENTER**. Enter the second value, **5**, and press **ENTER**. Continue this process and fill the matrix.

Press **2nd [Quit]**. To perform the test of independence, press **STAT**, highlight **TESTS**, and select **C: χ^2-Test** and press **ENTER**.

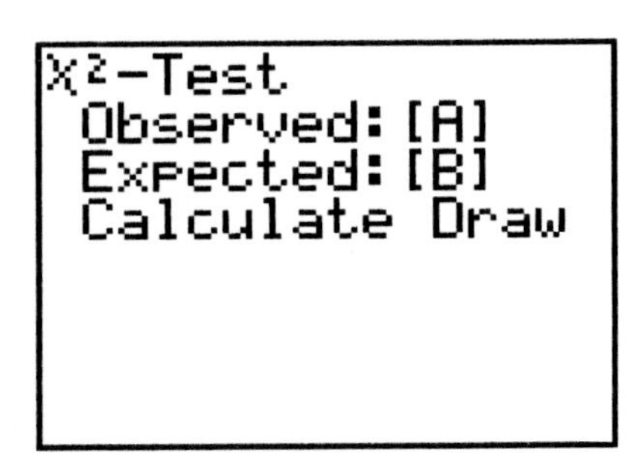

For Observed, [A] should be selected. If **[A]** is not already selected, press 2nd **[MATRX]**, highlight **NAMES**, select **1:[A]** and press ENTER. For, **Expected**, **[B]** should be selected. Move the cursor to the next line and select **Calculate** and press ENTER.

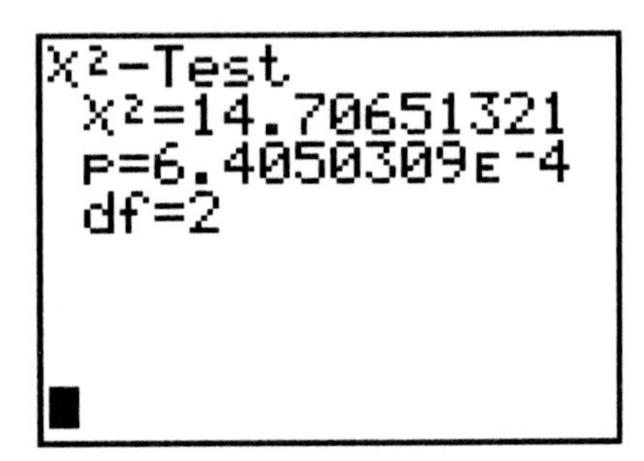

The output displays the test statistic and the P-value. Since the P-value is less than α, the correct decision is to **Reject** the null hypothesis. We conclude that at least one of the three groups experiences abdominal pain at a rate different from the other two groups.

The final step in this procedure is to confirm that the test requirements have been satisfied. The two requirements that we need to verify are: (1) all *expected frequencies* are greater than or equal to 1 and (2) no more than 20% of the *expected frequencies* are less than 5. Both requirements involve the *expected frequencies* which are stored in **Matrix B**. To view **Matrix B,** press **2nd [MATRX]** highlight **NAMES**, select **2:[B]** and press ENTER ENTER.

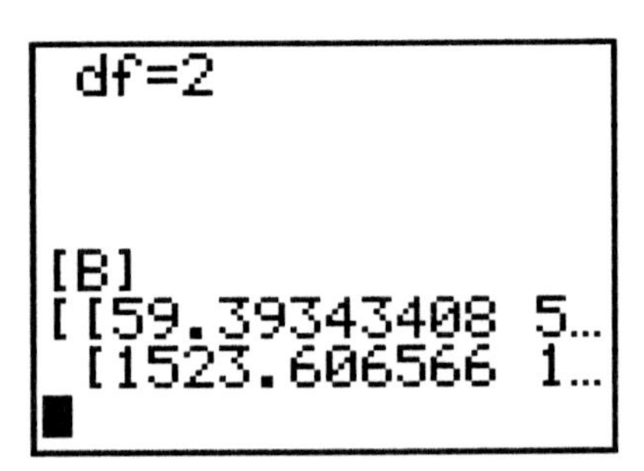

Scroll through the 6 entries in **Matrix B** and confirm that all entries are greater than 5. This confirms that the test requirements have been satisfied.

◀

Problem 8 (pg. 667)

In this example, we are testing the claim that amount of prenatal care is independent of wantedness of the pregnancy. The correct procedure is the Chi-Square (χ^2) Independence test. The hypotheses are: H_o : *Amount of prenatal care* is independent of *Wantedness of the pregnancy.* vs. H_a: *Amount of prenatal care* is not independent of *Wantedness of the pregnancy.*

Enter the data in the table into **Matrix A**.. Highlight **EDIT** and press **ENTER**. On the top row of the display, enter the size of the matrix. The matrix has 3 rows and 3 columns. Press **ENTER**. Fill the matrix with the data values.

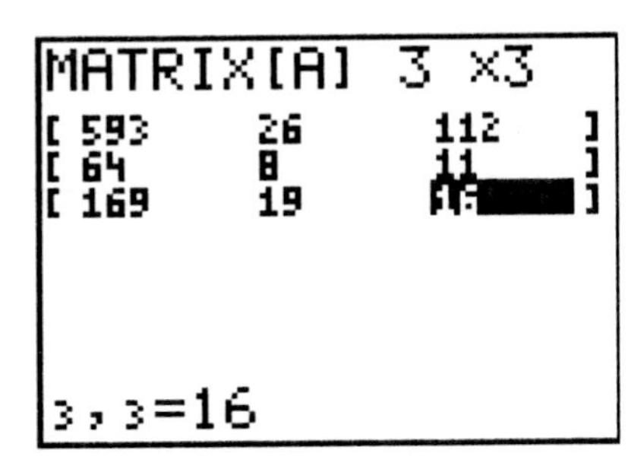

Press **2nd** **[Quit]**. To perform the test of independence, press **STAT**, highlight **TESTS**, and select **C: χ^2-Test** and press **ENTER**.

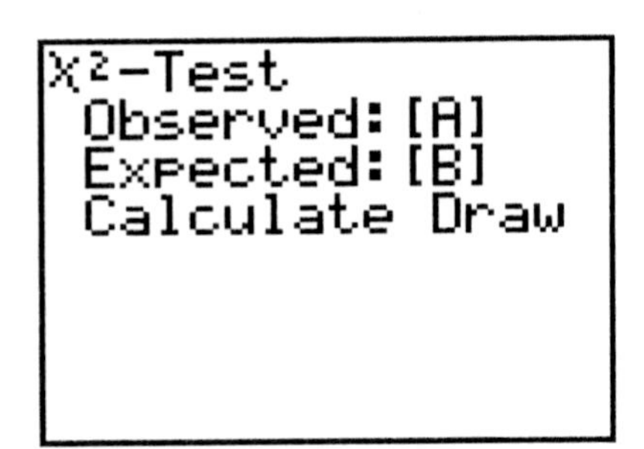

For **Observed**, **[A]** should be selected. If **[A]** is not already selected, press **2nd** **[MATRX]**, highlight **NAMES**, select **1:[A]** and press **ENTER**. For, **Expected**, **[B]** should be selected. Move the cursor to the next line and select **Calculate** and press **ENTER**.

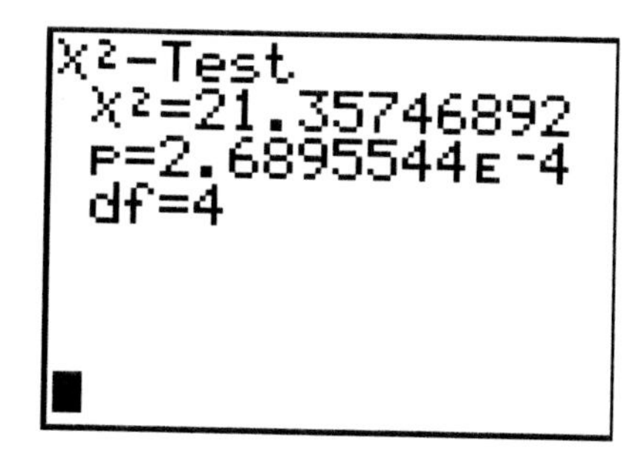

The output displays the test statistic and the P-value. Since the P-value is less than α, the correct decision is to **Reject** the null hypothesis. We conclude that *education level* and *region of the country* are **not** independent.

The final step in this procedure is to confirm that the test requirements have been satisfied. The two requirements that we need to verify are: (1) all *expected frequencies* are greater than or equal to 1, and (2) no more than 20% of the *expected frequencies* are less than 5. Both requirements involve the *expected frequencies* which are stored in **Matrix B**. To view **Matrix B,** press **2nd [MATRX]** highlight **NAMES**, select **2:[B]** and press **ENTER** **ENTER**.

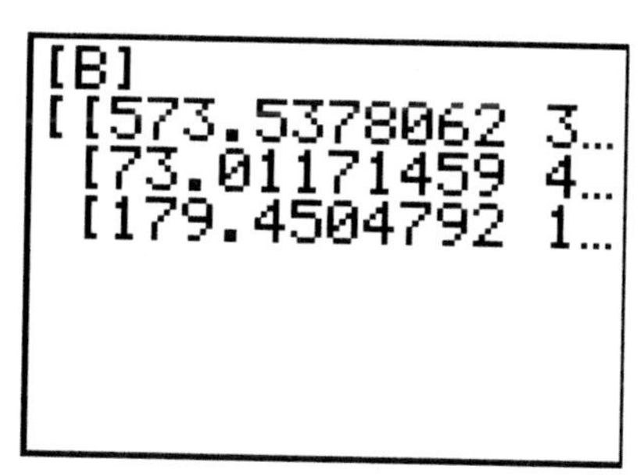

Scroll through the 9 entries in **Matrix B** (as you scroll through the entries, record these entries (rounded to the nearest whole number) on paper to use in part (b). of these problem) and confirm that all entries are greater than 5. This confirms that the test requirements have been satisfied.

(e).To determine which cell contributed most to the test statistic, we will store the two matrices into Lists.

Press **STAT**, highlight **EDIT.** Enter the *observed values* given in the Table into L1 (enter the 1st column, followed by the 2nd column, etc.) Enter the *expected values* from Matrix B (enter the 1st column, followed by the 2nd column, etc.) into L2. To calculate each cell's contribution to the χ^2-value, move the cursor so that it is flashing on '**L3**' at the top of the third column. Press **ENTER**. The cursor will move to the bottom of the screen and will be flashing next to '**L3=**'. Type in **(L1-L2)²/L2**. Press ENTER. Scroll through the values in L3 and find the largest value, **6.1**. This is the entry for 'More than 5 Months or Never' and 'Unintended.' The observed value for this cell is '11' and the expected value is '5.3'. This tells us that, for this cell, there were more observed women than we expected.

◀

TI-89 Instructions:

These instructions are designed to give you an overview of goodness of fit tests and tests for independence on the TI-89.

Goodness of Fit Test:
In the Stats/List Editor, enter the **observed frequencies** into **list1** and the **expected frequencies** into **list2**. Press 2nd F6 and select **7:Chi 2 GOF**. Use 2nd Var-Link to select **list1** as the **Observed List** and **list2** as the **Expected List**. The **Deg of Freedom** is equal to "**number of categories minus 1**." For **Results:** use the right arrow and select **Calculate**. Click ENTER until the output is displayed. The output contains the **Chi-Square** statistic, the **P-Value** and **df**. The **CompList** is a list of each individual cell's contribution to the Chi-Square statistic. If you press ENTER, you will see these individual values displayed in a list in the Stats/List editor.

Chi-Square Test for Independence:
Open the Data/Matrix Editor. Choose **3:New** to open a new matrix. On the next screen, on the **Type** line, use the right arrow to select **2:Matrix.** Move to the **Variable** entry box and type in a name for the matrix. Press ENTER. On the next two lines enter the row and column dimensions of the matrix. Press ENTER. Fill in the individual cell frequencies in the matrix. Press 2nd Quit. Open the Stats/List Editor. Press 2nd F6 and select **8:Chi 2 2-way.** Type in the name of your data matrix in the first entry box. Move to **Results:** and use the right arrow to select **Calculate.** Click ENTER until the output is displayed. The output includes the **Chi-Square** statistic, **P-Value, df** and a partial list of the **Expected cell values** and a partial list of the **individual cell contributions** to the Chi-Square statistic. The complete lists are stored in two matrices (**expmat** and **compmat**). To view these two matrices, return to the Data/Matrix Editor and select **2:Open**. Use the right arrow to select **2:Matrix** for Type. Press ENTER. Move to Folder and select **4:statvars.** Press ENTER. For variable, choose either **compmat** or **expmat**. Press ENTER twice to display the matrix that you selected.

TI-*n*spire Instructions:

These instructions are designed to give you an overview of goodness of fit tests and tests for independence on the **TI-*n*spire** handhelds.

Goodness of Fit Test:
Press (home) and select **6:New Document**. (Note: If you currently have a document open, the next screen will ask if you want to save the document. Press (tab) to select **No**. Press (enter).) Select **3:Add Lists & Spreadsheet.** Move to the top of column **A** and in the box next to '**A**' type in the name: **obs**. Move to **Line 1** in column **A** and begin entering the **observed values**. Move to the top of column **B** and in the box next to '**B**' type in the name: **expt**. Move to **Line 1** in column **B** and begin entering the **expected values**. Press (menu), select **4:Statistics, 4: Stat Tests** and select **7:** χ^2 **GOF.** On the next screen, use the down-arrow to select **obs** for the **Observed List.** Press (enter). Press (tab) to move to **Expected List** and use the down-arrow to select **expt.** Press (enter). Press (tab) to move to **df** and type in the degrees of freedom which is equal to "number of categories minus 1." Press (enter). Press (tab) to move to **1st Result Column** and type in 'C' for Column C. Press (enter). Press (tab) twice to move to **OK** and press (enter). The output is displayed in Columns **C** and **D**. The output includes the **Title** of the statistical test, χ^2 **statistic**, **P-Value** and **df.** The last entry is **"CompList."** It contains the **individual cell contributions** to the χ^2 statistic. To view the individual cell contributions, highlight the cell containing the values and press (enter). Use the right and left arrows to scroll through the values.

Chi-Square Test for Independence:
Press (home) and select **6:New Document**. (Note: If you currently have a document open, the next screen will ask if you want to save the document. Press (tab) to select **No**. Press (enter).) Select **1:Add Calculator.** Press (menu), select **7:Matrix & Vector,** select **6:Create** and select **1:New Matrix.** On the following screen, next to **newMat(** you need to enter the matrix dimensions. Suppose you have 3 columns of data with 4 rows per column, enter **4,3)** and press (enter). A **4x3 matrix** of zero's will be created. Use the up-arrow to highlight the matrix and press (enter). Use the left-arrow and the up-arrow to move to the first entry in the matrix. Press (clear) and enter the data value for that entry. Press the right-arrow twice, press (clear) and enter the data value for that entry. Continue moving through the matrix and replace all the zero's with the data values for each of the entries in the matrix. Once you have entered all the data, press (enter) to move outside the matrix, press (ctrl)(sto→ var) and type in a name for the matrix. Press (enter).

Press (menu), select **6:Statistics,** select **7:Stat Tests** and select **8:** χ^2 **2-way Test.** Press the down-arrow and select the matrix name. Press (enter). Press (tab) to highlight **OK** and press (enter). The output includes the **Title** of the statistical test, χ^2 **statistic**, **P-Value** and **df.** Two new matrices have been created. The **"Exp Matrix"** contains the **Expected cell values.** The **"Comp Matrix"** contains the **individual cell contributions** to the χ^2 statistic. To view these matrices, press (sto→ var) and select the matrix that you wish to view (s**tat.compmatrix** or **stat.expmatrix**) and press (enter).

Comparing Three Or More Means

Section 13.1

▸ Example 1 (pg. 681) Testing the Requirements of a One-Way ANOVA

The four requirements for the One-way ANOVA test are listed in your textbook on pg. 680. The first two requirements state that the data must be obtained using simple random sampling techniques and that all samples must be independent. We will assume that these requirements have been satisfied. The third requirement states that the populations must be normally distributed. This requirement can be validated through normal probability plots of each of the samples. The final requirement, that population variances are equal, can be tested using the sample standard deviations. The criteria that we will use is the following: the largest standard deviation must be no more than two times larger than the smallest standard deviation.

Enter the sample data from Table 1 on pg. 682 into **L1**, **L2, L3** and **L4**. We will check each dataset for normality using a normal probability plot. To set up the normal probability plot for the data in L1, press **2nd [STAT PLOT]**. Press **ENTER** to select **Plot 1**. Highlight **On** and press **ENTER**. Set **Type** to the normal probability plot which is the third selection in the second row. Press **ENTER**. Set **Data List** to **L1** and **Data Axis** to **X**. Next, there are three different types of **Marks** that you can select for the graph. The first choice, a small square, is the best one to use.

Press **ZOOM** and select **9:ZoomStat** and **ENTER**.

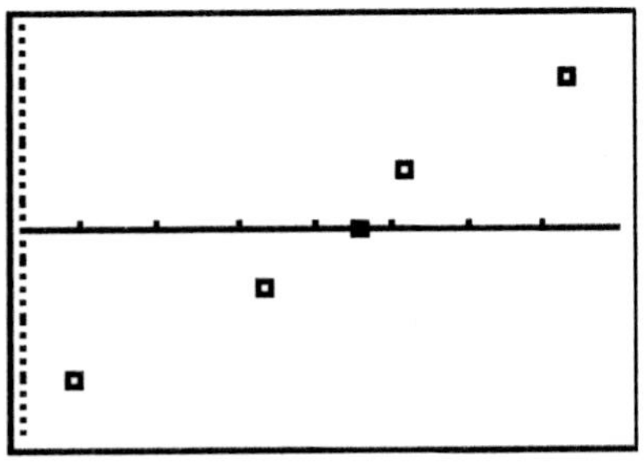

This plot is *fairly* linear, indicating that the data generally follow a normal distribution.

Repeat this process for the sample data in **L2, L3** and **L4**. All four plots are fairly normal so the requirement of normality has been satisfied.

The next requirement involves the sample standard deviations. To obtain the standard deviation of the data in **L1**, press **STAT**, select **CALC** and **1-Var Stats** and type in **2nd [L1.** Press **ENTER**. The standard deviation for the data in **L1** will appear on the screen (sx = 2.36). Repeat this process to obtain the standard deviation of the data in **L2** (2.60), **L3** (4.22) and **L4** (2.28). Calculate the ratio of the largest standard deviation to the smallest standard deviation: 4.22/2.28 = 1.851. Since this ratio is less than 2, the requirement of equal population variances is satisfied.

◀

▸ Example 3 (pg. 687) One-Way ANOVA Test

This example is a continuation of Example1.

The researcher wishes to determine if there is a difference in the mean shear bond strength among the four treatment groups. The test of hypothesis is:
$H_o : \mu_1 = \mu_2 = \mu_3 = \mu_4$ vs. H_a :at least one mean is different from the others.

In Example 1, we showed that the requirements of the One-Way ANOVA test have been satisfied. We also entered the data into **L1, L2, L3** and **L4**.

To run the hypothesis test, press STAT, highlight **TESTS** and select **F:ANOVA(** and type in **2nd [L1]** , **2nd [L2]** , **2nd [L3]** , **2nd [L4].**

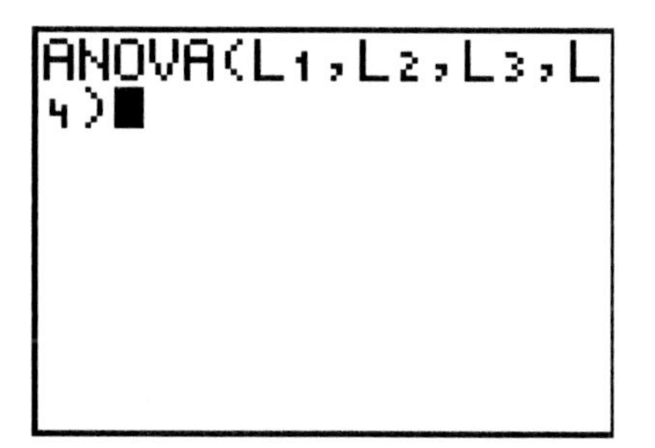

Press ENTER and the results will be displayed on the screen.

```
One-way ANOVA
 F=7.545199019
 p=.0022945915
 Factor
  df=3
  SS=199.9855
↓ MS=66.6618333
```

```
One-way ANOVA
↑ MS=66.6618333
 Error
  df=16
  SS=141.36
  MS=8.835
 Sxp=2.97237279
```

The output displays the test statistic, F = 7.545, and the P-value, p = 0.0023. Since the P-value is less than α, the correct decision is to **Reject** the null hypothesis. There is sufficient evidence to support the claim that at least one of the population means for shear bond strength among the four treatment groups is different.

The output also displays several other pieces of information. This information can be used to set up an Analysis of Variance Table (similar to the tables given as output in Minitab and Excel).

Note: The TI-84 can only do three boxplots on the same graph so it is not possible to get a graphical representation of the four data sets together.

◀

TI-89 Instructions:

These instructions are designed to give you an overview of ANOVA using the TI-89.

One-Way Analysis of Variance:
In the Stats/List Editor, if the raw data is given in the problem, enter the data into **list1, list2, list3,**... Press 2nd **F6** and select **C:ANOVA**. Select the Data Input Method: **Data** (if you have input the data into list1, list2, list3,...) or **Stats** (if you are using the summary statistics). Scroll to the **Number of Groups** and type in the appropriate number. Press ENTER. If you are using the raw data, use 2nd Var-Link to select the lists of data for **List1, List2, List3,**... The Analysis of Variance output displays the **F statistic** and the **P-Value**, followed by the various components of the ANOVA table (SS, MS, df, ...). Press ENTER and you will see three columns of calculated values: the mean for each data set and the lower and upper limits of the 95% Confidence Interval for the population mean for each data set.

Two-Way Analysis of Variance:
The **Two-Way ANOVA** can be accessed by pressing 2nd **F6** and selecting **D:ANOVA 2-Way.** Begin the analysis by entering the data into list1, list2, list3,... Select **D:ANOVA 2-Way.** Choose either the **Block** design or the **2 Factor, Eq Reps** design and fill in the required information.

TI-*n*spire Instructions:

These instructions are designed to give you an overview of ANOVA on the TI-*n*spire handheld.

One-Way Analysis of Variance:
Press (home) and select **6:New Document**. (Note: If you currently have a document open, the next screen will ask if you want to save the document. Press (tab) to select **No**. Press (enter).) Select **3:Add Lists & Spreadsheet** and enter either the data or the summary statistics for each group into separate columns in the spreadsheet. For example, if you are working with the summary statistics for three groups, assign names to Columns A, B and C. Enter into Column **A** the sample size, the mean and the standard deviation (in that order) for Group 1. Enter the Group 2 information into Column **B** and enter the Group 3 information into Column **C**.

Press (menu), select **4:Statistics, 4:Stat Tests** and **C:ANOVA.** Select the Data Input Method: **Data** (if you have input the actual data into the spreadsheet) or **Stats** (if you have input the summary statistics). Tab to the entry: **Number of Groups** and type in the number of groups. Press (tab) to highlight **OK** and press (enter). For **Group 1 Stats,** use the down-arrow and select the name for column A. Press (enter). Press (tab) to highlight **Group 2 Stats** and use the down-arrow and select the name for column B, etc. For the entry: **1st Result Column**, type in the letter of the next available column that has not been used for data. Press (tab) to highlight **OK** and press (enter).

The Analysis of Variance output displays the **F statistic** and the **P-Value**, followed by the various components of the ANOVA table (SS, MS, df, …). Following the ANOVA table information, there are three more entries. The first entry contains the lower limits of the 95% Confidence Interval for the population mean for each Group. The second entry contains the upper limits of the 95% Confidence Interval for the population mean for each Group. The third entry contains the sample means for each Group.

Two-Way Analysis of Variance:
The **Two-Way ANOVA** can be accessed by pressing (menu) and selecting **4:Statistics, 4:Stat Tests** and **D:ANOVA 2-Way.** Begin the analysis by entering the data into the spreadsheet. Choose either the **Block** design or the **2 Factor, Eq Reps** design and fill in the required information.

Inferences on the Least Squares Regression Model

Section 14.1

▸ Example 1 (pg. 737) Least-Squares Regression

Press **STAT**, highlight **1:Edit** and clear **L1** and **L2**. For each of the fourteen patients, enter the age into **L1** and the total cholesterol into **L2**. Press **2nd [STAT PLOT]**, select **1:Plot1**, turn **ON** Plot 1 and press **ENTER**. For **Type** of graph, select the **scatter plot** which is the first selection. Press **ENTER**. Enter **L1** for **Xlist** and **L2** for **Ylist**. Highlight the first selection, the small square, for the type of **Mark**. Press **ENTER**. Press **ZOOM** and **9** to select **ZoomStat**.

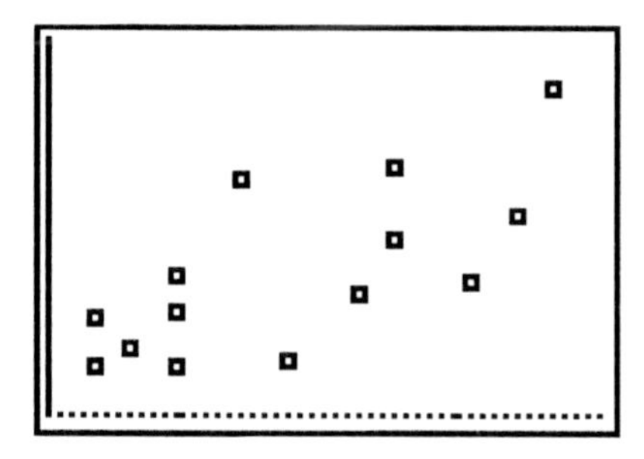

This graph shows a positive linear correlation, with quite a bit of scatter.

In order to calculate r, the correlation coefficient, and, r^2, the coefficient of determination, you must turn **On** the **Diagnostic** command. Press **2nd [CATALOG]** (Note: **CATALOG** is found above the **0** key). The CATALOG of functions will appear on the screen. Use the down arrow to scroll to the **DiagnosticOn** command.

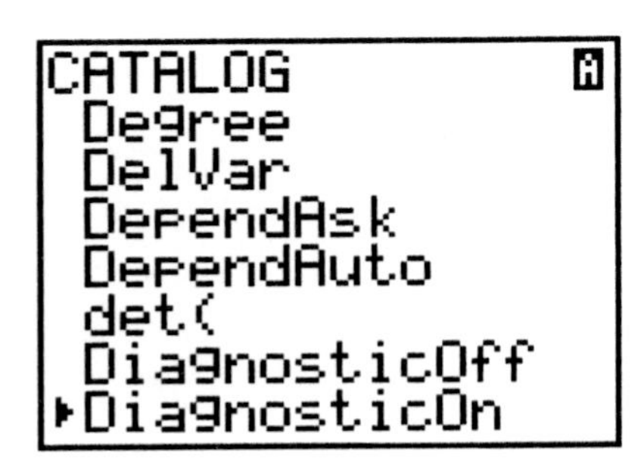

Press **ENTER** **ENTER**.

To calculate the correlation coefficient, the coefficient of determination and the regression equation, press **STAT**, highlight **CALC**, select **4:LinReg(ax+b).** Press **VARS**, highlight **Y-VARS**, select **1:Function** by pressing **ENTER** , select **1:Y1** by pressing **ENTER** and press **ENTER**. This stores the regression equation in **Y1.** (Note: This command gives you the option of specifying which lists contain the X-values and Y-values. If you do not specify these lists, the defaults are used. The defaults are: **L1** for the X-values and **L2** for the Y-values.)

```
LinReg
 y=ax+b
 a=1.399064152
 b=151.3536582
 r²=.5152520915
 r=.7178106237
```

The correlation coefficient is r = .718. This suggests a positive linear correlation between X and Y, but not a very strong one. The coefficient of determination is .515. This tells us that 51.5% of the variation in cholesterol levels can be explained by the predictor variable, age. The regression equation is $\hat{y} = 151.35 + 1.399x$.

To see a scatterplot of the data along with the regression equation, press **GRAPH**

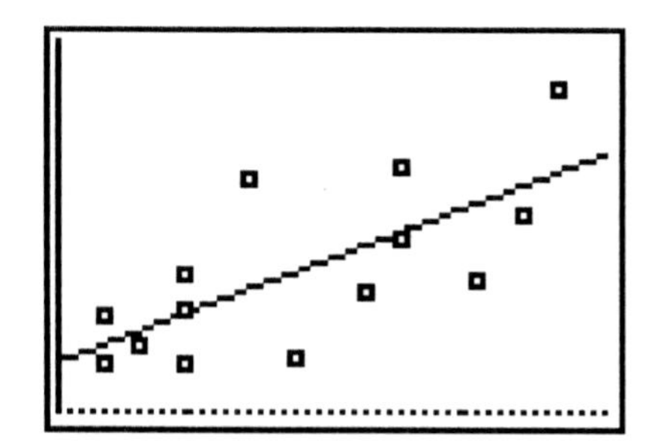

◀

▸ Example 3 (pg. 741) Computing the Standard Error

This example is a continuation of the previous one. (Note: If you have not already done so, enter the data from Table 1 on pg.737 into **L1** and **L2**. Press **STAT**, highlight **CALC**, select **4:LinReg(ax+b)**. Press **VARS**, highlight **Y-VARS**, select **1:Function** by pressing **ENTER**, select **1:Y1** by pressing **ENTER** and press **ENTER**. (This stores the regression equation in **Y1.**)

Press **STAT**, highlight **TESTS** and select **E:LinRegTTest**. Enter **L1** for **Xlist**, **L2** for **Ylist**, and **1** for **Freq**. On the next line, β and ρ, select $\neq 0$ and press **ENTER**. Leave the next line, RegEQ, blank. Highlight **Calculate**.

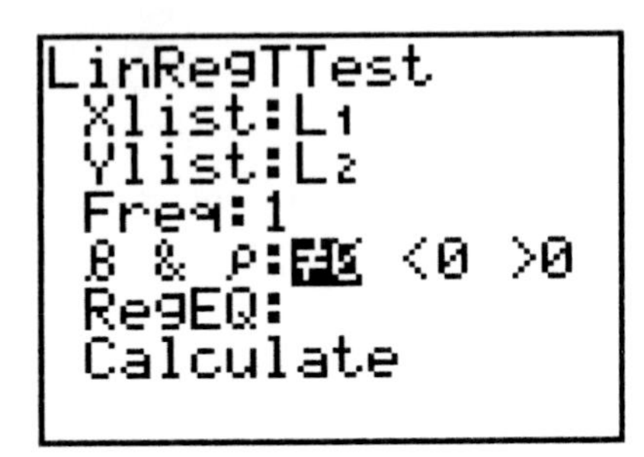

Press **ENTER**.

```
LinRegTTest
 y=a+bx
 β>0 and ρ>0
 t=3.57143321
 p=.0019211307
 df=12
↓a=151.3536582
```

The output displays several pieces of information describing the relationship between X and Y. What you are interested in for this example is the standard error. Scroll down to the next page of output and you will see s = 19.48. This is the standard error.

◂

Example 4 (pg. 742) Verifying that the Residuals are Normally Distributed

This example is a continuation of the previous ones. (Note: If you have not already done so, enter the data from Table 1 on pg.737 into **L1** and **L2**. Press **STAT**, highlight **CALC**, select **4:LinReg(ax+b)**. Press **VARS**, highlight **Y-VARS**, select **1:Function** by pressing **ENTER**, select **1:Y1** by pressing **ENTER** and press **ENTER**. (This stores the regression equation in **Y1.**)

The values for $(y_i - \hat{y}_i)$, called Residuals, are automatically stored to a list called **RESID**. Press **2nd [LIST]**, select **7:RESID**. Press **STO**, **2nd [L3]** and **ENTER**. This stores the residuals to **L3**.

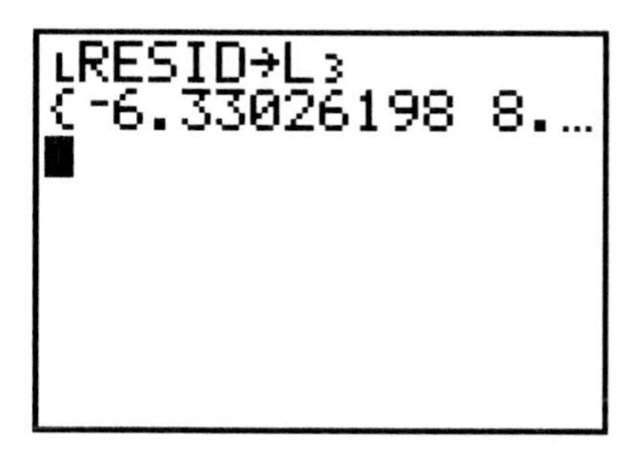

To set up the normal probability plot, press **2nd [STAT PLOT]**. Press **ENTER** to select **Plot 1**. Highlight **On** and press **ENTER**. Set **Type** to the normal probability plot which is the third selection in the second row. Press **ENTER**. Set **Data List** to **L3** and **Data Axis** to **X**. For **Marks** select the small square.

Press **ZOOM** and select **9:ZoomStat** and **ENTER**.

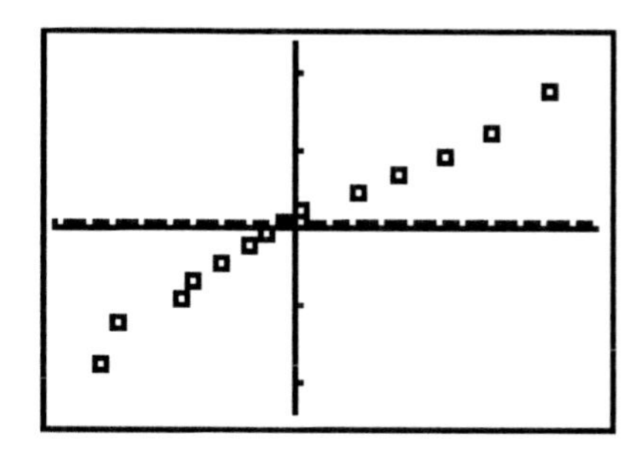

This plot is *fairly* linear, indicating that the Residuals follow a normal distribution.

Example 6 (pg. 745) Testing for a Linear Relation

This example is a continuation of the previous examples. (Note: If you have not already done so, enter the data from Table 1 on pg.737 into **L1** and **L2**. Press **STAT**, highlight **CALC**, select **4:LinReg(ax+b)**. Press **VARS**, highlight **Y-VARS**, select **1:Function** by pressing **ENTER**, select **1:Y1** by pressing **ENTER** and press **ENTER**. This stores the regression equation in **Y1.)**

To test the claim that there is a linear relationship between *age* and *total cholesterol,* the appropriate hypothesis test is: $\beta_1 = 0$ vs. $\beta_1 \neq 0$.

The first step is to verify that the assumptions required to perform the test are satisfied. The first assumption is that the sample has been obtained using random sampling. This has been confirmed and is stated in Example 1. The next assumption is that the residuals are normally distributed and this has been confirmed by the normal probability plot in Example 4. The last assumption is that the residuals have constant error variance. This assumption can be validated using a graph of the Residuals vs. the predictor variable, Age. If the Residuals do, in fact, have constant error variance, then the Residuals will appear as a scatter of points about the horizontal line at 0.

To construct the graph of the Residuals vs. Age, press **2nd [STAT PLOT]**. Press **ENTER** to select **Plot 1**. Highlight **On** and press **ENTER**. Set **Type** to the scatter plot which is the first selection in the first row. Press **ENTER**. Set **Xlist** to **L1** and **YList** to **L3**. For **Marks,** select the small square.

Press **WINDOW** and use the data to set the Window. First, look at the *Age* variable in Table 1 on pg. 737. Notice the minimum and maximum values (25 and 65). Choose **Xmin** and **Xmax** to encompass these values. For example, choose **Xmin** = 20 and **Xmax** = 70. To set the Y-values, you must look through the Residuals that are stored in **L3**. Press **2nd [QUIT]**. Press **STAT**, highlight **EDIT** and scroll through the values in **L3**. The minimum value is –24.5 and the maximum value is 34.482. To make the graph symmetric about the X-axis, use **Ymin** = -35 and **Ymax** = 35. Press **WINDOW** and enter these values for **Ymin** and **Ymax**. Press **GRAPH**

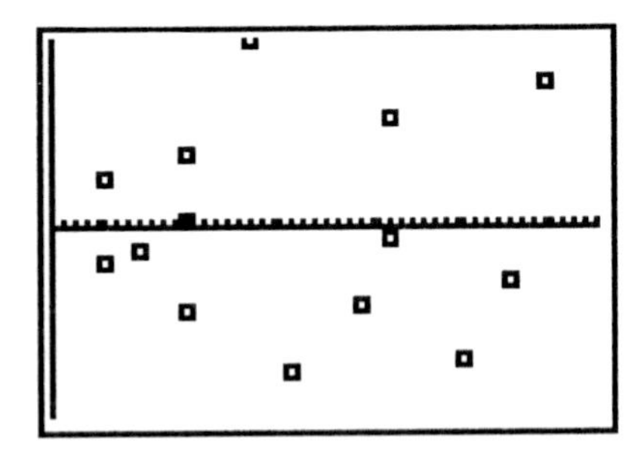

The errors are evenly spread about the horizontal line at 0, so the assumption of constant error variance is satisfied.

To run the test, press STAT, highlight **TESTS** and select **E:LinRegTTest**. Enter **L1** for **Xlist**, **L2** for **Ylist**, and **1** for **Freq**. On the next line, β and ρ, select $\neq 0$ and press ENTER. Leave the next line, RegEQ, blank. Highlight **Calculate**.

```
LinRegTTest
 Xlist:L1
 Ylist:L2
 Freq:1
 β & ρ:≠0 <0 >0
 RegEQ:
 Calculate
```

Press ENTER.

```
LinRegTTest
 y=a+bx
 β≠0 and ρ≠0
 t=3.57143321
 p=.0038422614
 df=12
↓a=151.3536582
```

```
LinRegTTest
 y=a+bx
 β≠0 and ρ≠0
↑b=1.399064152
 s=19.48053511
 r²=.5152520915
 r=.7178106237
```

The output displays several pieces of information describing the relationship between X and Y. What you are interested in for this example is the P-value (p = .0038). Since the P-value is less than α, the correct decision is to **Reject** the null hypothesis. This indicates that there is a linear relationship between X and Y. ◀

▸ Example 7 (pg. 747) Constructing a Confidence Interval about the Slope of the True Regression Line

This example is a continuation of the previous examples. (Note: If you have not already done so, enter the data from Table 1 on pg.737 into **L1** and **L2**. Press **STAT**, highlight **CALC**, select **4:LinReg(ax+b)**. Press **VARS**, highlight **Y-VARS**, select **1:Function** by pressing **ENTER**, select **1:Y1** by pressing **ENTER** and press **ENTER**. This stores the regression equation in **Y1.)**

The 95% confidence interval for β_1, the slope of the true regression line is given by the following formula:

$$b_1 \pm t_{\frac{\alpha}{2}} \cdot \frac{s_e}{\sqrt{\sum (x_i - \bar{x})^2}}.$$

From the previous examples in this Section, we have already obtained the following values: b_1 and s_e. b_1 is the coefficient of x in the regression equation and s_e is the standard error. The value for $t_{\frac{\alpha}{2}}$ can be found using the TI-84. (Note: The TI-83 does not have this option.) To find the t-value on the TI-84, you will need a value for $\frac{\alpha}{2}$ and also, a value for the degrees of freedom. Since we are constructing a 95% confidence interval for β_1, α = 5% or .05. So, $\frac{\alpha}{2}$ = .025. The degrees of freedom value = (n – 2) = (14 – 2) = 12. To find the t-value, press **2nd DISTR** and select **4:invT**. This command requires two values, $\frac{\alpha}{2}$ and degrees of freedom. Type in **.025** , **12**) and press **ENTER**.

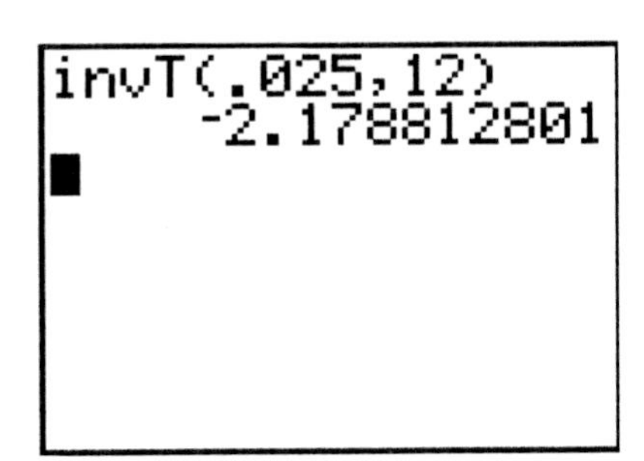

The t-value for the lower bound of the confidence interval is – 2.179. The corresponding t-value for the upper bound of the confidence interval is 2.179.

The only part of the formula that we now need to calculate is the denominator and we can use an equivalent form of this expression that can be easily evaluated on the calculator. The denominator, $\sqrt{\sum(x_i - \bar{x})^2}$, can be written equivalently as: $\sqrt{(\sum x_i^2 - \frac{(\sum x_i)^2}{n}}$. The values for $\sum x_i and \sum x_i^2$ are stored in your calculator as part of the regression procedure. To access them, press VARS and select **5:Statistics.** On the Statistics Menu, select Σ. Then select the first entry, Σx by pressing ENTER ENTER. The result, 589, will appear on the screen. To obtain $\sum x_i^2$, press VARS and select **5:Statistics.** On the Statistics Menu, select Σ. Then select the second entry, Σx^2, by pressing ENTER ENTER. The result is 27253. Enter these values into the equation:

$$\sqrt{(27253 - \frac{(589)^2}{14}} = 49.73.$$

Now fill the values into the formula for the confidence interval.
The lower bound of the 95% confidence interval is:

$$1.399 - 2.179.\frac{19.48}{49.73} = 0.545.$$

The upper bound is:

$$1.399 + 2.179.\frac{19.48}{49.73} = 2.253.$$

◀

Problem 16 (pg. 749)

In this exercise, we will go directly to the **LinRegTTest**. This procedure combines all the steps for analyzing the model. It has more output for analyzing the regression model then the command **LinReg(ax+b).**

To begin the analysis, enter the data into L1 and L2. Next, press STAT, highlight **TESTS** and select **E:LinRegTTest**. Enter **L1** for **Xlist**, **L2** for **Ylist**, and **1** for **Freq**. On the next line, β and ρ, select $\neq 0$ and press ENTER. Move the cursor to the next line, RegEQ. On this line, tell the calculator where to store the regression equation. We will store it in **Y1.** Press VARS, highlight **Y-VARS**, select **1:Function** by pressing ENTER, select **1:Y1** by pressing ENTER. Highlight **Calculate**.

```
LinRegTTest
 Xlist:L1
 Ylist:L2
 Freq:1
 β & ρ:≠0 <0 >0
 RegEQ:Y1
 Calculate
```

Press ENTER.

```
LinRegTTest
 y=a+bx
 β≠0 and ρ≠0
 t=12.50191637
 p=7.6259241E-8
 df=11
↓a=.2088364861
```

```
LinRegTTest
 y=a+bx
 β≠0 and ρ≠0
↑b=.057527857
 s=.1121198008
 r²=.9342490303
 r=.9665655851
```

(a.) First, notice the regression equation: y= a+bx. The unbiased estimator of β_0 is 'a' which is equal to 0.2088. The unbiased estimator of β_1 is 'b' which is equal to 0.0575.
(b.) The standard error is 's' which is equal to 0.1121.
(c.) The values for $(y_i - \hat{y}_i)$, called Residuals, are automatically stored to a list called **RESID**. Press **2nd** **[LIST]**, select **7:RESID**. Press **STO**, **2nd** **[L3]** and **ENTER**. This stores the residuals to **L3**.

To set up the normal probability plot, you must first 'deselect' Y1. Press **Y=** and move the cursor so that it highlights **=** in the Y1 equation and press **ENTER**. This will deselect Y1 so that it will not appear on the graph with the Residual plot. Next, press **2nd** **[STAT PLOT]**, select **1:Plot1**, turn **ON** Plot 1 and press **ENTER**. Set **Type** to the normal probability plot which is the third selection in the second row. Press **ENTER**. Set **Data List** to **L3** and **Data Axis** to **X**. For **Marks** select the small square.

Press **ZOOM** and select **9:ZoomStat** and **ENTER**.

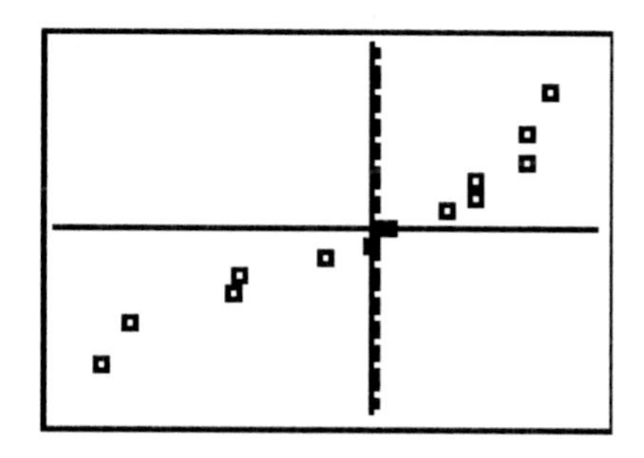

This plot is *fairly* linear, indicating that the Residuals follow a normal distribution.

(d.) To calculate s_{b_1}, we will use the formula $\frac{s_e}{\sqrt{\sum (x_i - \overline{x})^2}}$. The denominator, $\sqrt{\sum (x_i - \overline{x})^2}$, can be written equivalently as:

$\sqrt{(\sum x_i^2 - \frac{(\sum x_i)^2}{n}}$. The values for $\sum x_i and \sum x_i^2$ are stored in your calculator as part of the regression procedure. To access them, press **VARS** and select **5:Statistics.** On the Statistics Menu, select Σ. Then select the first entry, Σx by pressing **ENTER** and press **ENTER**. The result, 184, will appear on the screen. To obtain $\sum x_i^2$, press **VARS** and select **5:Statistics.** On the Statistics Menu, select Σ. Then select the second entry, Σx^2, by pressing **ENTER** **ENTER**. The result is 3198. Enter these values into the equation:

$$\frac{.1121}{\sqrt{3198-\frac{(184)^2}{13}}}=.0046$$

(e.) To test the claim that there is a linear relationship between the predictor variable, 'Tar' and the response variable, 'Nicotine', the appropriate hypothesis test is: $\beta_1 = 0$ vs. $\beta_1 \neq 0$. This is the test that we set up in the **LinRegTTest**. Notice that a p-value is displayed in the output. Since the p-value of 7.626E-8 (written in scientific notation) is less than the α-value of .10, the correct decision is to **reject H_0.** The data supports the claim that there is a linear relationship between the variables.

(f.) The 90% confidence interval for β_1 is: $b_1 \pm t_{\frac{\alpha}{2}} s_{b_1}$. To calculate the confidence interval, replace all the variables with their appropriate values (b_1 = .0575, s_{b_1} = .0046). To find the t-value on the TI-84, you will need a value for $\frac{\alpha}{2}$ and also, a value for the degrees of freedom. Since we are constructing a 90% confidence interval for β_1, α = 10% or .10. So, $\frac{\alpha}{2}$ = .05. The degrees of freedom value = (n – 2) = (13 – 2) = 11. To find the t-value, press **2nd DISTR** and select **4:invT**. This command requires two values, $\frac{\alpha}{2}$ and degrees of freedom. Type in **.05 , 11)** and press **ENTER** to obtain the t-value. The confidence interval for β_1 is: .0575 ± 1.796(.0046).

(g.) To obtain the mean amount of nicotine in a cigarette that has 12 mg of tar, press **VARS**, highlight **Y-VARS**, select **1:Function** by pressing **ENTER**, select **1:Y1** by pressing **ENTER** Type in (12) and press **ENTER** **Y1(12) = 0.899.**

Section 14.2

▸ Example 3 (pg. 756) Constructing a Confidence Interval and a Prediction Interval

This example is a continuation of the Examples in Section 14.1

To construct a confidence interval for the mean y-value at a specific x-value, we must obtain the value for $\hat{y}$ from the regression equation and then calculate the margin of error. The confidence interval is then created using the formula $\hat{y}$ ± margin of error. The formula that we will use for the margin of error is equivalent to the one in the textbook. The formula we will use is:

$$E = t_{\frac{\alpha}{2}} s_e \sqrt{\left(\frac{1}{n} + \frac{n(x^* - \bar{x})^2}{n(\sum x_i^2) - (\sum x_i)^2}\right)}$$

In the following steps we will find all the values needed to construct the confidence interval.

First, start with the regression equation that we obtained in Example 1: $\hat{y} = 151.35 + 1.399x$. Using this equation we calculate $\hat{y}$ for x = 42 and obtain the value, 210.1.

Next, we will find the t-value on the TI-84. We will need a value for $\frac{\alpha}{2}$ and also, a value for the degrees of freedom. Since we are constructing a 95% confidence interval, α = 5% or .05. So, $\frac{\alpha}{2}$ = .025. The degrees of freedom value = (n – 2) = (14 – 2) = 12. To find the t-value, press **2nd DISTR** and select **4:invT**. This command requires two values, $\frac{\alpha}{2}$ and degrees of freedom. Type in **.025** **,** **12** **)** and press **ENTER**.

Next, we need the standard error, s_e, which we obtained in Example 3 in Sec. 14.1. The value is 19.48.

Next, we need to calculate $\bar{x}$, $\sum x^2$, and $(\sum x)^2$. Press **VARS**, select **5:Statistics**. Highlight **2:** $\bar{x}$ and press **ENTER** **ENTER**. Notice that $\bar{x}$ = 42.07.

Press **VARS** again, select **5:Statistics,** highlight $\sum$, select **1:** $\sum x$ and press **ENTER** **ENTER**. So, $\sum x = 589$. Press **VARS** again, select **5:Statistics,** highlight $\sum$, select **2:** $\sum x^2$, and press **ENTER** **ENTER**. Notice that $\sum x^2 =$ 27253.

Now, calculate the margin of error, E, when $x^* = 42$ using the formula for E shown above.

```
2.179*19.48*√(1/
14+(14(42-42.07)
²/(14((27253)-(5
89)^2))
         11.34441534
```

The confidence interval for the mean y-value at x = 42 is $\hat{y} \pm 11.34$. Since $\hat{y} = 210.1$, we calculate 210.1 ± 11.34 to obtain the lower bound of 198.76 and the upper bound of 221.44. These are the lower and upper bounds of the 95% confidence interval for the mean cholesterol level of all 42-year-old females.

To construct a prediction interval for the y-value at a specific x-value, we will follow the same type of procedure as we did for the confidence interval: $\hat{y} \pm$ margin of error. The only change is in the formula for the margin of error. The formula that we will use for the margin of error is equivalent to the one in the textbook. The formula we will use is:

$$E = t_{\frac{\alpha}{2}} s_e \sqrt{(1 + \frac{1}{n} + \frac{n(x^* - \bar{x})^2}{n(\sum x_i^2) - (\sum x_i)^2}}$$

Calculate the margin of error, E, when $x^* = 42$ using the formula for E shown above.

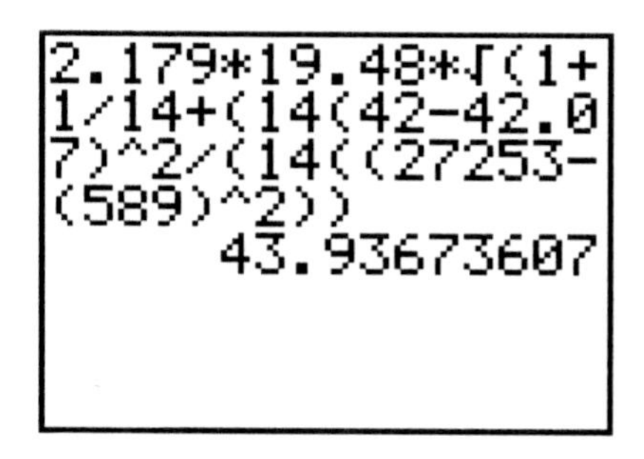

The prediction interval for the y-value at x = 42 is $\hat{y} \pm 43.94$. Since $\hat{y} = 210.1$, we calculate 210.1 ± 43.94 to obtain the lower bound of 166.16 and the upper bound of 254.04. These are the lower and upper bounds of the 95% prediction interval for the cholesterol level of a randomly selected 42-year-old female.

◀

TI-89 Instructions:

These instructions are designed to give you an overview of the Linear Regression T-test using the TI-89.

Linear Regression Test:
In the Stats/List Editor, enter the x-values into list1 and the y-values into list2. To analyze the linear relationship between X and Y, press **2nd** **F6** and select **A:LinRegTTest**. Use **2nd** **Var-Link** and select **list1** for **X List**, and select **list2** for **Y List**. For Alternate Hyp select **β & ρ ≠ 0**. **Store RegEqn** to **y1(x)**. For Results: select **Calculate** and press **ENTER**.

The output displays:

The **t-statistic** and **P-value** for the alternate hypothesis: **β & ρ ≠ 0;**

the degrees of freedom, **df;**

the values for **a** and **b** in the linear equation: y = a + bx;

the standard error, s;

the standard error of the slope, s_b: $\dfrac{s_e}{\sqrt{\sum (x_i - \bar{x})^2}}$;

the coefficient of determination: r^2;

and the linear correlation, r.

To construct a 95% Confidence interval for β, use the formula: $b \pm t_{\frac{\alpha}{2}} s_b$. The values for b and s_b, have been calculated in the LinReg TTest. To find a value for $t_{\frac{\alpha}{2}}$, we need a value for $\dfrac{\alpha}{2}$ and also, a value for the degrees of freedom.

Since we are constructing a 95% confidence interval, α = 5% or .05. So, $\dfrac{\alpha}{2}$ = .025. The degrees of freedom value = (n – 2). To find the t-value, in the Stats/List editor, press **F5** and select **2:inverse** and select **2:Inverse t** and press **ENTER**. On the next screen, enter **.025** for **Area** and the value of **n - 2** for **Deg of Freedom**. Press **ENTER** twice. The t-value is displayed in the output screen.

TI-*n*spire Instructions:

These instructions are designed to give you an overview of the Linear Regression t-test using the TI-*n*spire handhelds.

Press (home) and select **6:New Document**. (Note: If you currently have a document open, the next screen will ask if you want to save the document. Press (tab) to select **No**. Press (enter).) Select **3:Add Lists & Spreadsheet.** Move to the top of column **A** and in the box next to '**A**' type in a name. Move to **Line 1** in column **A** and begin entering your **x-values**. Move to the top of column **B** and in the box next to '**B**' type in a name. Move to **Line 1** in column **B** and begin entering your **y-values**. Press (menu), select **4:Statistics,** select **4: Stat Tests** and select **A:Linear Reg t Test.** On the next screen, press the down arrow and highlight the name you are using for column **A**. Press (enter). Press (tab) to move to **YList**. Use the down arrow and highlight the name you are using for Column **B**. Press (enter). Press (tab). You can use the default name to '**Save RegEqn**.' Tab to the entry: **Alternate Hypothesis** and choose the alternate hypothesis (most often '≠' is the correct selection.) Tab to the entry: 1st **Result Column** and type in '**c**' for Column **C**. Press (tab) to highlight **OK** and press (enter). The regression output is displayed in Columns **C** and **D**.

The output displays:

The Title of the statistical test,

the alternate hypothesis: $\rho \neq 0$,

the form of the regression equation: y = a + bx,

t-statistic, P-value and df,

the values for **a** and **b** in the linear equation: y = a + bx;

the standard error, s;

the standard error of the slope, SE slope: $\dfrac{s_e}{\sqrt{\sum (x_i - \bar{x})^2}}$;

the coefficient of determination: r^2;

the linear correlation: r,

and the Residuals.

To view the residuals highlight the cell containing the residuals and press (enter). Use the right and left arrows to scroll through the residual values.

To construct a 95% Confidence interval for β, use the formula: $b \pm t_{\frac{\alpha}{2}} s_b$. The values for b and s_b, have been calculated in the LinReg tTest. press (home) and select **1:Calculator.** To find a value for $t_{\frac{\alpha}{2}}$, we need a value for $\frac{\alpha}{2}$ and also, a value for the degrees of freedom. Since we are constructing a 95% confidence interval, α = 5% or .05. So, $\frac{\alpha}{2}$ = .025. The degrees of freedom value = (n – 2). To find the t-value, press (menu), select **5:Probability,** select **5:Distributions** and **6:Inverse t.** Type **.025** for **Area** and press (enter). Press (tab) and enter the value of **n - 2** for **Deg of Freedom**. Press (tab) to highlight **OK** and press (enter). The t-value is displayed in the output screen.